Cellular Nanomachines

Bhanu P. Jena

Cellular Nanomachines

From Discovery to Structure-Function and Therapeutic Applications

Springer

Bhanu P. Jena
Department of Physiology
Wayne State University School of Medicine
Detroit, MI, USA

ISBN 978-3-030-44495-2 ISBN 978-3-030-44496-9 (eBook)
https://doi.org/10.1007/978-3-030-44496-9

This Springer imprint is published by the registered company Springer Nature Switzerland AG
The registered company address is: Gewerbestrasse 11, 6330 Cham, Switzerland

To Minakhi and Siddhartha
You are my inspiration

Foreword

Major discoveries and inventions are made by the prepared mind. While human curiosity drives creativity, urging us to explore our natural world, it necessitates and nurtures the invention and development of new tools and approaches to explore. Reciprocally, new discoveries also lead to new inventions. In the past 400 years since the invention of the light microscope leading to the discovery of the unit of life the cell, there has been an explosion in our understanding of the chemistry of life. This new understanding has taught us that Nature is the ultimate architect and engineer, capable of designing and building the most effective and efficient living unit the "Cell" and the precision nanomachines within, to undertake various life functions. Cellular nanomachines are a marvel of Nature, undertaking with great precision the vital cellular tasks like synthesis of new proteins life's building blocks to their proper folding and assembly, their transport, and their secretion from within the cell to the outside. Secretion is required for cell–cell communication such as neurotransmission for coordination, thought, memory, smell, vision, taste, and hearing, for the digestion of food, for endocrine control such as insulin release in response to elevated blood glucose, and for release of immune products in response to a foreign antigen such as a pathogenic bacteria or virus, among others. In this book, the author lucidly takes the reader on a journey of the major paradigm-shifting discoveries made in the past century on key cellular nanomachines, their complex yet precise and elegant design and function, the diseases linked to their dysfunction, and the therapeutic approaches to overcome them. The major focus of this book is on the "porosome" nanomachine, the universal secretory portal in cells. The discovery of the porosome was made using atomic force microscopy at the nanoscale resolution of live cells. Each of these discoveries is pioneering and outstanding scientific contributions of great service to humanity.

Iowa State University
Ames, IA, USA

Lloyd L. Anderson,

Preface

The unit of life is the cell. An average human body contains nearly 38 trillion cells, equal to a linear length that could circle half the globe. The cell is analogous to a well-organized city, with thousands of reactions and millions of interactions occurring at any given moment to sustain life processes. A fundamental understanding of these vital cellular processes therefore is critical to comprehending life and in the detection and treatment of diseases that develop as a consequence of altered states. Since the discovery of the cell in 1665 using the then newly invented light microscope, great progress has been made in our understanding of its structure and function using a wide range of imaging modalities. In the past century, new tools and approaches, especially multimodal imaging, have enabled the discovery of cellular nanomachines, providing a molecular-level understanding of cellular structure–function. The average size of a cell is about 15 μm. In comparison, the thickness of a human hair is 150 μm. Cellular nanomachines, Nature's engineered marvels, are the life-sustaining cellular workhorses. These nanomachines measure just 15–150 nm; hence, 10,000 of 15 nanometer-sized structures could fit in the thickness of a human hair. In this book written for the general readership, the discovery of these life-sustaining cellular nanomachines, a brief description of their structure–function, and their impact on health and disease are discussed. Except for one, most of the cellular nanomachines discussed in this book include schematic illustrations for simplicity and clarity.

Among the key cellular nanomachines discovered are the "*porosome complex*" or "secretory nanomachine" at the cell plasma membrane for energy-dependent transient docking, fusion, and the selective and precise fractional release of intravesicular contents from cells during secretion; the "*nuclear pore complex*" for the precise and selective bidirectional transport of proteins and nucleotides between the nucleus and the cell cytosol; the energy-dependent protein-folding machinery or "*chaperonin*"; "*proteasome*," the energy-dependent garbage disposal in cells; "*ribosome*," the protein synthetic machinery in cells; "*ATP synthase*," the nanomachine that generates cellular energy; and "*myosin*," the cellular molecular motor for movement and transport.

The earliest of cellular nanomachines discovered was the molecular motor protein *myosin* in 1859 by Wilhelm Kühne [Kühne, W. (1859) *Arch. f. Anat. Physiol. u. wissensch. Med.* 748]. Next was the discovery of the *nuclear pore* complex in 1949 by Harold G. Callan using electron microscopy [Callan H. G., Randall J. T., Tomlin S. G. (1949). An electron microscope study of the nuclear membrane *Nature* 163, 280]. In 1955, the *ribosome* was discovered by George E. Palade also using transmission electron microscopy [Palade, G.E. (January 1955). A small particulate component of the cytoplasm. *J Biophys Biochem Cytol.* 1 (1): 59–68.]. In 1958, Efraim Racker discovered *ATP synthase* [Pullman ME, Penefsky H, Racker E. 1958. A soluble protein fraction required for coupling phosphorylation to oxidation in submitochondrial fragments of beef heart mitochondria. *Arch. Biochem. Biophys.* 76:227] and was the first to identify in the 1960s the F1-ATPase with the knobs seen in electron micrographs of mitochondrial membranes. The concept of the presence of a protein-folding machinery later termed *chaperonin*, distinct from the known ability of proteins to self-fold, was born out of the work of Costa P. Georgopoulos in 1972 [Georgopoulos, C. P., Hendrix, R. W., Kaiser, A. D., Wood, W. B. 1972. Role of the host cell in bacteriophage morphogenesis: effects of a bacterial mutation on T4 head assembly. *Nat. New Biol.* 239:38]. Similarly, the *proteasome* was discovered in 1979 by Avram Hershko, Aaron Ciechanover, and Irwin A. Rose [Hershko, A.; Ciechanover, A.; Rose, I.A. (1979), "Resolution of the ATP-dependent proteolytic system from reticulocytes: a component that interacts with ATP," *Proc. Natl. Acad. Sci. USA*, 76 (7): 3107]. The secretory machinery *porosome* was discovered in my laboratory in 1996 (published online) using atomic force microscopy [Schneider, S.W., Sritharan, K.C., Geibel, J.P., Oberleithner, H., Jena, B.P. 1997. Surface dynamics in living acinar cells imaged by atomic force microscopy: Identification of plasma membrane structures involved in exocytosis. *Proc. Natl. Acad. Sci. USA.* 94:316] and later imaged using electron microscopy and solution X-ray.

In the subsequent years following their discoveries, each cellular nanomachine has been intensely investigated by many laboratories who have made major contributions to a greater understanding of their structure using multiple imaging modalities, their function and composition, their functional reconstitution, and an understanding of their involvement in a number of diseases associated with their dysfunction. The final chapter explores some of the operating principles such as chirality and enthalpy that may govern the assembly of higher-order structures such as cellular nanomachines.

While a large body of information in the form of in-depth reviews and textbook chapters are available on cellular nanomachines discussed in this book, seldom are readers exposed to the scientific journey leading to their discovery. It is the objective of this book to provide the reader with the background, time-line, and approaches that lead to the discovery and understanding of each of the seven cellular nanomachines. The author intends readers to be critically aware that all discoveries are made possible due to the accumulated body of scientific information, hypothesis, and tools developed and available to the inquisitive and prepared mind. The contents of this book are therefore far from exhaustive, rather it provides the reader with a brief narration of the scientific journey leading to the discovery of each cellular nanomachines and their structure–function and participation in health and disease. The "porosome" though discussed in various textbooks is the most recently (1996–1997) discovered nanomachine and therefore is discussed in greater detail. The author thanks Dr. Won Jin Cho for the modified schematic illustration presented in Figs. 2.1, 3.1, 4.1, 5.1, 6.1, and 7.1 and cover illustration.

Detroit, MI, USA
December 25, 2019

Bhanu P. Jena

Introduction

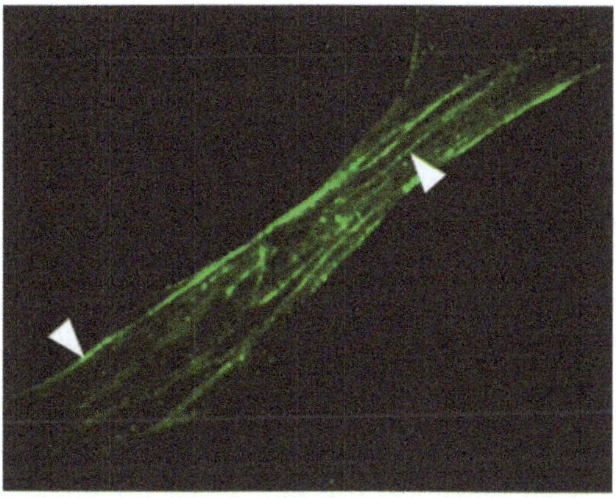

Left: Arrangement of myosin IIb fibers in a human skeletal muscle cell in culture (*From: ACS Biomater. Sci. Eng. 2019, 5, 9*). ©Bhanu Jena

Cellular nanomachines are Nature's engineered marvels that perform with great efficiency and precision the life-sustaining cellular activities. *Myosin*, for example, is the universal molecular motor in living organisms and belongs to a superfamily of motor proteins. Analogous to a car engine that utilizes chemical energy in the form of gasoline to perform mechanical work to propel the vehicle forward, the myosin molecular motor uses chemical energy in the form of adenosine triphosphate or ATP for locomotion. Myosin typically has a globular head region (heavy chain) measuring 16.5 nm in length and 6.5 nm in width and a tail (superhelical rod) measuring up to 150 nm in length. The myosin head has ATP binding and ATPase activity that utilizes ATP as fuel to perform mechanical work, leading to a wide range of motions, from transport of subcellular organelles within cells to cellular motion and to the locomotion of an entire organism via skeletal muscle myosin II. Myosin II is a 520 kDa hexamer comprised of two heavy chains and four light chains. Each heavy chain is 220 kDa, composed of a short globular head and a long coiled-coil tail region. The two myosin II light chains modulating force transduction have a molecular mass of 22 and 19 kDa, respectively, and belong to a large family of calcium-binding proteins.

In contrast, the nanomachine *ATP synthase* is specialized with the task of generating the universal life-sustaining energy currency ATP in all living organisms. ATP synthase measures approximately 10 nm in diameter and 20 nm in height and has two regions F0 and F1, with each region composed of several subunits. The F0 region is embedded in the inner mitochondrial membrane and serves as a proton pore. The F0 subunits are arranged in a ring embedded in the mitochondrial inner membrane, and when an F0 subunit is protonated, it undergoes a conformational change that drives neighboring subunits to rotate, resulting in conformational changes in the F1 subunits and consequent catalysis of adenosine diphosphate (ADP) and Pi to ATP. Analogous to myosin, the other cellular nanomachines like the ribosome, the nuclear pore, porosome, proteasome, and chaperonin, all require ATP to operate, demonstrating the critical role of ATP synthase.

Ribosomes are approximately 30 nm in size protein synthetic nanomachines that are themselves composed of nearly 80 different proteins. The ribosome is composed of two subunits: a small subunit which reads the mRNA and the large subunit which precisely links specific amino acids to form a polypeptide chain. During polypeptide biosynthesis at the endoplasmic reticulum, linear polypeptide chains undergo self-folding to form 3D mature and functional proteins, while some require assistance. This assistance in the folding of nascent linear polypeptides is provided by a large class of cylindrical nanomachines called *chaperonins*, each measuring approximately 15 3nm, where ATP-dependent folding of polypeptides occurs in addition to the prevention of their aggregation.

Misfolded or damaged proteins, and certain proteins that are required to be degraded, are carried out by another unique cellular nanomachine called the *proteasome*. The proteasome is a 15 × 11.5 nm cylindrical structure with a 1.3 nm pore through which only a partially folded or linear polypeptide can enter for ATP-mediated proteolysis. The proteasome has been referred to as the garbage disposal of the cell.

Similarly, communication between cells in a multicellular organism is critical in maintaining homeostasis. This is achieved via chemical messages such as the hormone insulin that is secreted from beta cells of the endocrine pancreas to maintain blood glucose levels, or secretion of neurotransmitters from a brain neuron to communicate with other neurons in the process for thought, learning, or memory. Secretion of digestive enzymes from acinar cells of the exocrine pancreas following food intake is required for digestion. The cellular nanomachine at the cell plasma membrane that accomplishes this vital cellular task of secretion is the *porosome*. Porosomes are unidirectional cup-shaped lipoprotein secretory nanomachines ranging in size from 15 nm at the nerve terminal to 180 nm in the exocrine pancreas.

While porosomes are composed of 30–40 proteins, the 120 nm *nuclear pore complex* is a bidirectional transporting nanomachine, composed of nearly 1000 protein molecules. The nuclear pore complexes are aqueous channels traversing the outer and inner nuclear membrane, serving as gatekeepers for the selective transport of proteins and nucleotides between the nucleoplasm and the cytoplasm. Although ions can freely move through the nuclear pore, proteins, mRNA, ribosome subunits, and tRNA are actively transported through the complex. *In this Springer Nature book, the porosome nanomachine will be the primary focus.*

Contents

Chapter 1
Porosome: Cells Secretory Nanomachine

Abstract Secretion is a highly regulated fundamental cellular process in living organisms, from yeast to cells in humans. Cellular cargoes such as neurotransmitters in neurons, insulin in beta cells of the endocrine pancreas, or digestive enzymes in the exocrine pancreas are all packaged and stored in membrane-bound secretory vesicles that dock and fuse at the cell plasma membrane to release their contents during secretion. The prevailing view was that secretory vesicles completely merge with the cell plasma membrane, emptying the entire vesicular contents outside the cell during secretion. However, accumulation of partially empty secretory vesicles observed in electron micrographs in cells following a secretory episode suggested fractional release of intra-vesicular contents during cell secretion. Given the high surface tension at the secretory vesicle membrane, fractional intra-vesicular content release during cell secretion could only be possible via a plasma membrane structure capable of preventing the complete merger or collapse of secretory vesicles into the cell plasma membrane. Cup-shaped plasma membrane-embedded lipoprotein structure called ***porosomes*** was first discovered in 1996 in live pancreatic acinar cells using atomic force microscopy (AFM) and subsequently confirmed in all cells examined including neurons using AFM, electron microscopy (EM), and solution X-ray. The porosome exhibits dynamics and its chemical composition demonstrates the utilization of energy in the form of both ATP and guanosine triphosphate (GTP), the participation of molecular motors, ion channels, and soluble N-ethylmaleimide-sensitive factor activating protein receptor (SNARE) membrane fusion proteins, among others. Porosomes are composed of nearly 30 proteins, as opposed to the 120 nm nuclear pore complex comprised of nearly 1000 protein molecules. Porosomes range in size from 15 nm in neurons and astrocytes to 100–180 nm in endocrine and exocrine cells. Porosome has been functionally reconstituted into artificial lipid membrane and in live cells. During secretion, secretory vesicles dock at the base of the porosome complex via v-SNARE proteins at the secretory vesicle membrane and t-SNARE proteins at the porosome base. In the presence of calcium, the v-SNARE and t-SNARE proteins in the opposing bilayers interact in a circular array to establish conducting channels or ***fusion pores***. An

B. P. Jena, *Cellular Nanomachines*,
https://doi.org/10.1007/978-3-030-44496-9_1

increase in volume of the docked secretory vesicle via the rapid entry of ions and aquaporin-mediated rapid entry of water molecules results in increased intra-vesicular pressure, enabling the fractional release of vesicular contents from the cell with great precision. Collectively, these observations provide a molecular understanding of the fractional release of intra-vesicular contents via the ***transient*** or ***kiss-and-run*** mechanism of cell secretion. The discovery of the porosome and the molecular mechanism of its structure–function has resulted in a paradigm shift in our understanding of the secretory process in cells.

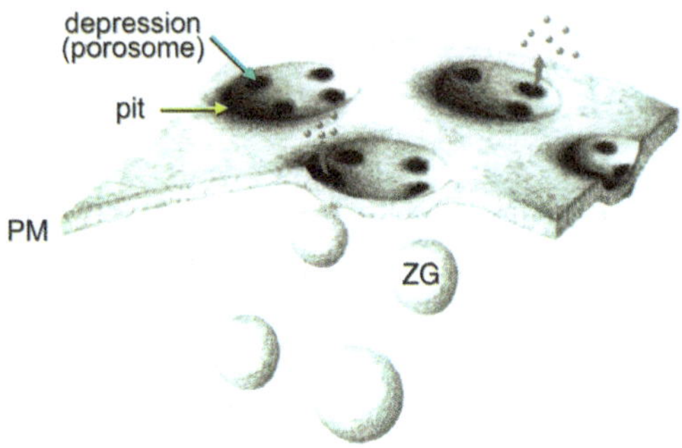

Keywords Cell secretion · Porosome · Secretory portal · Fractional vesicle content release

Introduction

Secretion is a fundamental process through which cells communicate with their environment and exchange information in a multicellular context to reach homeostasis and sustain life. For decades, the prevailing view was that secretion operates as an all-or-none "complete fusion" event, where vesicles are trafficked to the cell surface where they fuse and completely incorporate into the plasma membrane. Vesicle contents then diffuse out of the cell and the excess membrane incorporated into the cell plasma membrane is retrieved by compensatory endocytosis. This hypothesis, although attractive at first glance, had several key setbacks. First, following a secretory episode, partially empty secretory vesicles accumulate within cells as observed in electron micrographs, demonstrating that secretory vesicles are capable of partial content release during secretion. Second, it predicted a quantization of secretory products packaged into each secretory vesicle, when in fact, secretory vesicle size greatly varies even within the same cell, sometimes as much as sixfold. Third, the level of additional regulation necessary to rapidly and precisely internalize and sequester vesicle-associated lipids and proteins following incorporation into the cell plasma membrane seems extraordinarily complex, given the tens of thousands of different membrane lipids and their differential distribution even between opposing leaflets in the same bilayer. We therefore hypothesized the presence of a tunable dial at the cell plasma membrane for transient secretory vesicle docking and fusion (via t-SNARE and v-SNARE proteins arranged in a rosette or ring complex) without full collapse, and the generation of hydrostatic pressure within vesicles to drive with great precision, intra-vesicular contents to the cell exterior. Through the use of atomic force microscopy, we identified nanoscale transmembrane cup-shaped lipoprotein structures and named them "porosomes." In the past 24 years, porosomes have been implicated in a wide range of secretory events [1] (Fig. 1.1). The family of proteins that constitute the porosome have been biochemically characterized and the mesoscale structure of the complex has been well documented using AFM, EM, and solution X-ray methods. Defects in one or more porosome components have measurable, often highly potent effects on the regulation of secretion, establishing

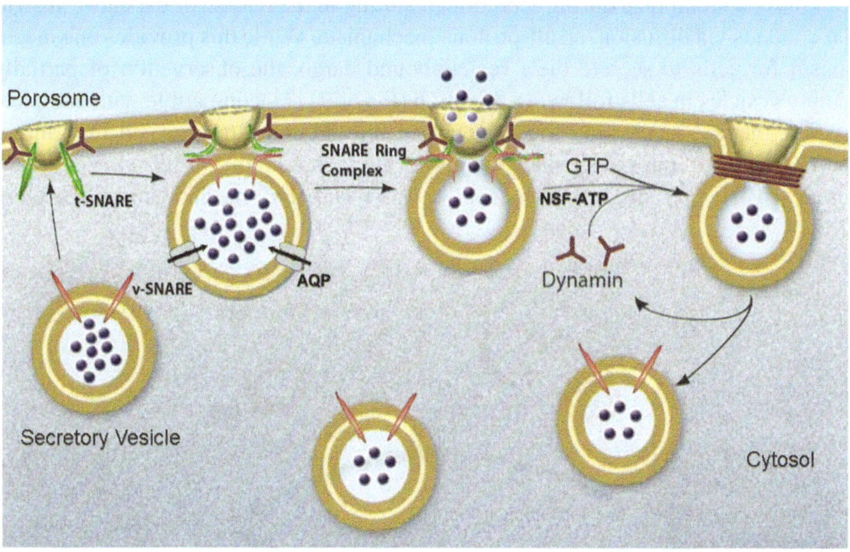

Fig. 1.1 Schematic drawing depicting porosome-mediated fractional release of intra-vesicular contents from the cell during secretion. Secretory vesicles dock at the porosome base in the cell plasma membrane via t-SNAREs present at the porosome base and v-SNAREs present at the secretory vesicle membrane, to establish a ***t-/v-SNARE ring complex***. In the presence of calcium, the SNARE ring complex promotes the establishment of continuity between the opposing bilayers ***[fusion pore]*** through which pressurized intra-vesicular contents [intra-vesicular pressure established via active transport of water through aquaporin or water channels (AQP) at the secretory vesicle membrane] are expelled to the outside during cell secretion. On completion of secretion, the t-/v-SNARE rosette or ring complex is disassembled by the ATPase *N*-ethylmaleimide sensitive factor (NSF) and the fused lipid membrane is cleaved by the GTPase dynamin. The resultant partially empty secretory vesicle dissociates from the porosome base and is endocytosed. In exocrine and endocrine cells, the partially empty vesicle may undergo multiple rounds of exo-endocytosis prior to degradation and recycling. However, in fast secretory cells such as the neuron, neurotransmitter transporters at the synaptic vesicle membrane rapidly replenish the vesicle with neurotransmitters from the cytosol, rendering it secretory competent. ©Bhanu Jena

links between point mutations and secretion-defective disease states such as cystic fibrosis that were previously correlative and are now causative [2, 3]. The discovery of the porosome solved the conundrum of fractional discharge of intra-vesicular contents from cells by providing an explanation for regulated graded secretion. Porosomes are cup-shaped supramolecular lipoprotein structures at the cell plasma membrane ranging in size from 15 nm in neurons to 100–180 nm in endocrine and exocrine cells and composed of approximately 30 proteins. In comparison, the 120 nm nuclear pore complex is composed of nearly 1000 protein molecules.

Porosome Discovery

Since the 1950s, it was believed that secretory vesicles completely merge with the cell plasma membrane during secretion, resulting in the release of the entire vesicular contents via diffusion, an all-or-none mechanism. While this provides one mechanism for cells to secrete their vesicle-bound cargo, the observation of partially empty vesicles in cells following secretion (Fig. 1.2) is incompatible with complete vesicle merger, suggesting the presence of an additional mechanism that allows partial release of intra-vesicular contents during secretion. In 1973 *Bruno Ceccarelli* proposed the presence of "transient" or "*kiss-and-run*" mechanism of secretory

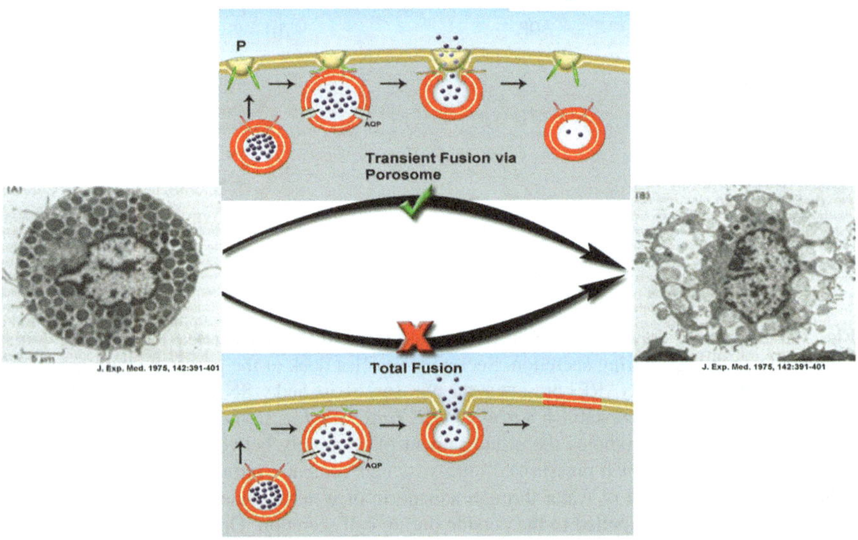

Fig. 1.2 Fractional release of intra-vesicular contents could only be achieved via *porosome*-mediated transient fusion and release mechanism. Electron micrographs of rat peritoneal mast cells, in resting (**a**, extreme left) and following secretion (**b**, extreme right). Note the fractional release of intra-vesicular contents following secretion (**b**) (*J. Exp. Med.* 1975, 142:391–401). This fractional release of intra-vesicular contents could only be achieved via the *porosome* (P)-mediated transient fusion and release mechanism shown (√). ©Bhanu Jena

vesicle fusion at the cell plasma membrane enabling fractional release of intra-vesicular contents from the cell during secretion [4]. In his 1973 study [4] using frog nerve-muscle preparation, *Ceccarelli* reported that after a period of 4 h *"of stimulation, the terminals were severely depleted of transmitter, but not of vesicles, and their general morphological organization was normal."* In 1990 *Wolfhard Almers* hypothesized that the *fusion pore*, a continuity established between the vesicle membrane and the cell plasma membrane, may result from a "preassembled ion channel-like structure that could open and close" [5]. Later in a 1992 review [6], *Julio Fernandez* opined that the difficulty in observing such channel-like structures was the lack of ultrahigh-resolution imaging tools to directly monitor their presence and study their dynamics in live cells. *"It seems terribly wasteful that, during the release of hormones and neurotransmitters from a cell, the membrane of a vesicle should merge with the plasma membrane to be retrieved for recycling only seconds or minutes later."* wrote *Erwin Neher* in 1993 [7].

In the mid-1990s, there were frequent discussions in my laboratory that since secretory vesicles have high curvature, they would fuse and collapse at the cell plasma membrane. However, to prevent such vesicle collapse and merger, there needs to be a "*structure at the cell plasma membrane*" that would help guide the secretory vesicle to the structure and promote transient vesicle docking and fusion to enable partial content release. Furthermore, to precisely release a portion of intra-vesicular contents during cell secretion, there has to exist a measured regulated increase in intra-vesicular pressure generated from the entry of water and ions into the docked vesicle. The intra-vesicular pressure generated could then drive a portion of the intra-vesicular contents through the transiently established fusion pore at the specialized plasma membrane structure. Such a mechanism we hypothesized would enable the release of a measured portion of the intra-vesicular contents, prior to vesicle disengagement (endocytose) from the cell plasma membrane, thereby retaining the identity of both the secretory vesicle membrane and the cell plasma membrane (Fig. 1.2 a (√), b).

To test this hypothesis and motivated by the goal to identify cellular structures at the plasma membrane involved in the regulated fractional release of intra-vesicular contents from cells during secretion, in 1994, we employed the then new nanoscale imaging tool—the atomic force microscope (AFM). The morphology and dynamics of nanoscale surface structures in live pancreatic acinar cells were examined in real time in resting and following stimulation of cell secretion, using the AFM. The pancreatic acinar cell was chosen as a model system for the proposed study, primarily due to its slow secretory dynamics and for being a well characterized and extensively studied secretory cell. In pancreatic acinar cells, secretion occurs in the minute timescale, compared to seconds and milliseconds, respectively, in neuroendocrine cells and neurons. Following optimization of the AFM for live cell imaging, the major breakthrough came in 1995–1996, when circular pit-like structures containing 100–180 nm in diameter and 40–100 nm depressions or pores (Fig. 1.3a, b) were observed for the first time at the apical plasma membrane (where secretion is known to occur) in live pancreatic acinar cells. On stimulation of secretion, the depressions grew larger, returning to their resting size following completion of

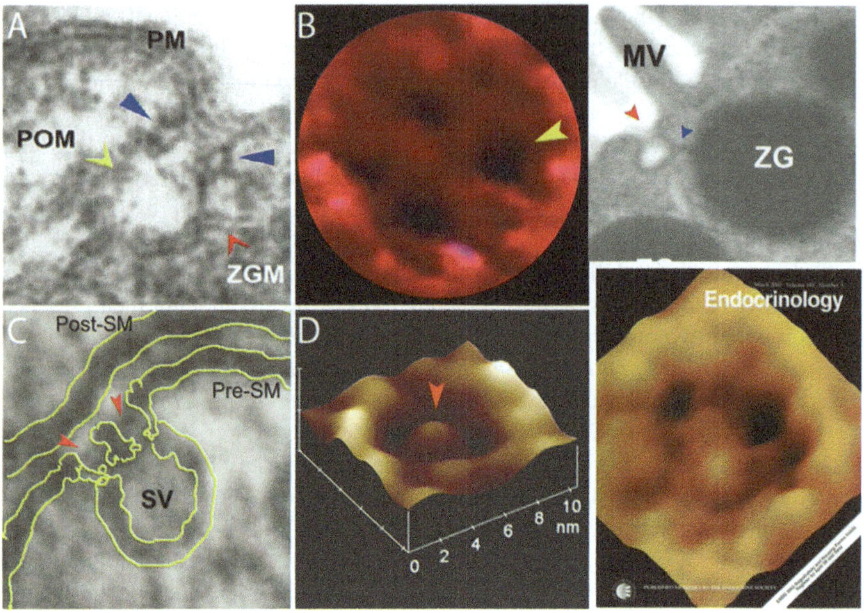

Fig. 1.3 Porosomes are universal secretory portals at the cell plasma membrane. Porosomes in acinar cells of the exocrine pancreas (**a** [EM], **b** [AFM] and EM micrograph: Top Right), neurons (**c** [EM], **d** [AFM]), growth hormone secreting cell of the pituitary gland (Journal cover: Bottom Right). (**a**) Electron micrograph of a single *porosome* at the apical plasma membrane (PM) of a pancreatic acinar cell showing the *porosome* membrane (POM, yellow arrowhead) associated with the membrane of a secretory vesicle called zymogen granule (ZGM). Cross section of a circular ring structure (blue arrowhead) possibly composed of actin–myosin forms the neck of the *porosome* complex involved in the dilation of the *porosome* opening following stimulation of secretion. (**b**) AFM micrograph of the apical surface topology of a live pancreatic acinar cell demonstrating the presence of four *porosomes* (one indicated by the yellow arrowhead) opening to the cell's exterior. *Porosomes* in the exocrine pancreas range in size from 100 to 180 nm in diameter. (**c**) EM micrograph of a neuronal *porosome* (red arrowheads) with a docked synaptic vesicle (SV) at its base, in the presynaptic membrane (Pre-SM) of the nerve terminal. Note the central plug in the neuronal *porosome* complex. *Porosomes* in neurons range in size from 10 to 17 nm in diameter. (**d**) AFM micrograph of a neuronal *porosome* at the presynaptic membrane in an isolated synaptosome. Note the central plug (red arrowhead). (*Extreme Top Right*) EM micrograph of *porosome* (Fig. 1.3 top left and right) next to microvilli (MV) at the apical plasma membrane (PM) of a pancreatic acinar cell with docked secretory vesicle or ZG. (*Extreme Lower Right*) AFM micrograph of the apical surface topology of a live GH cell from pig pituitary, demonstrating the presence of 100–180 nm in diameter *porosomes* (black circular openings). [Images obtained from: *Proc Natl Acad Sci* 94:316–321 (1997); *Biophys J* 85:2035–2043 (2003); *Cell Biol Int* 28:699–708 (2004); *J Microscopy* 232:106–111 (2008); *Endocrinology* 143:1144–1148 (2003)]. ©Bhanu Jena

secretion. Exposure of acinar cells to actin depolymerizing mycotoxin cytochalasin D, a known inhibitor of cell secretion, resulted in the collapse of depressions and a loss of amylase secretion. *These results identify for the first time new cellular structures at the cell plasma membrane involved in secretion* [8]. Five years later, new results from our laboratory established that the observed depressions are the long predicted secretory portals at the cell plasma membrane [9, 10]. In January 2002 and February 2003, we reported in two separate studies [9, 10] that following stimulation of cell secretion, gold-conjugated polyclonal monospecific amylase-specific antibodies accumulate at depressions, establishing depressions to be secretory portals at the cell plasma membrane [8–10]. The starch-digesting enzyme amylase is a major intra-vesicular product secreted by the exocrine pancreas. The term "*porosome*" was later assigned to these pores or depression structure. The reported study [9] additionally demonstrated the presence of t-SNAREs at the *porosome* base facing the cytosol, firmly establishing *porosome* structures to be secretory portals at the cell plasma membrane where secretory vesicles dock and fuse via v-SNAREs at the vesicle membrane with t-SNAREs at the porosome base. Electron micrographs of *porosome* at the apical plasma membrane (PM) of pancreatic acinar cells with docked secretory vesicle called zymogen granule (ZG) are shown in Fig. 1.3 (Fig. 1.3 top left and right). In Fig. 1.3a, the *porosome* membrane (POM, yellow arrowhead) associated with the ZG membrane (ZGM) is shown. In the same year in a separate study, we reported *porosome* structure and dynamics in chromaffin cells [11] and then in March of 2002, our laboratory in collaboration with *Lloyd Anderson* reported [12] (Fig. 1.3 cover illustration lower right) the structure of *porosome* and their dynamics at the cell plasma membrane in growth hormone (GH) secreting cells of the pig pituitary gland and the accumulation of GH-immuno-gold at *porosome* openings following stimulation of cell secretion. These results further demonstrate *porosome* to be plasma membrane-associated secretory portals in cells. In September 2003 [13] following isolation of the *porosome* from acinar cells of the exocrine pancreas and determination of its composition and reconstitution into artificial lipid membrane, the *porosome* was recognized as the universal secretory portal in cells [13]. In this study [13], morphological details of the *porosome* complex associated with docked secretory vesicles were revealed using EM (Fig. 1.3a). Finally, in 2016, the insulin-secreting *porosome* was isolated, characterized, and functionally reconstituted in live insulin-secreting beta cells [14], establishing its critical role as the secretory nanomachine in cells and its promise for future therapeutic applications.

Porosome Structure and Dynamics

Porosomes are cup-shaped lipoprotein structures present at the cell plasma membrane, measuring 10–17 nm in neurons and astrocytes and from 100 nm to 180 nm in endocrine and exocrine cells. High-resolution imaging of live cells at the nanometer scale using AFM has provided in great detail the native structure and dynamics of the porosome complex in live cells. At the apical region of pancreatic acinar cells where secretion occurs, the plasma membrane surface is decorated with 500–2000 nm in diameter and 45 nm deep pits. On average, pits measure 400 nm with four *porosomes* present within each pit. *Porosomes* in acinar cells of the exocrine pancreas measure 100–180 nm in diameter and 40–100 nm in depth (Figs. 1.4, 1.5, and 1.6).

Exposure of pancreatic acinar cells to a secretory stimulus results in the dilation of the porosome openings that return to its resting size following completion of secretion (Fig. 1.6) [8]. Five minutes following a secretory stimulus, the diameter of the pancreatic porosome opening increases by 35% and a corresponding 129% increase in amylase secretion is observed (Fig. 1.6) [8]. Exposure of pancreatic acinar cells to cytochalasin D, a fungal toxin that inhibits actin polymerization and known to inhibit cellular secretion, results in a 15–20% drop in porosome diameter and a 50–60% loss in the stimulated secretion of the starch-digesting enzyme amylase [8]. Results from these studies were the first to suggest porosomes (depressions at the cell surface) to be the secretory portals in pancreatic acinar cells. Furthermore, these studies demonstrated the involvement of actin in the regulation of both the structure and function of porosomes.

Similar to pancreatic acinar cells, porosomes in resting growth hormone (GH) secreting cells of the pituitary gland measure 150 nm in diameter [12]. Following a secretory stimulus, a 40% increase in the GH cell porosome opening diameter and its return to resting size following completion of secretion has been reported [12]. Furthermore, the loss of GH secretion following exposure to the actin depolymerizing agents cytochalasin D has also been reported in GH cells [12], implicating the porosome as the secretory portal in GH cells. Similarly, porosomes in insulin-secreting β-cells of the endocrine pancreas [15], neurons [16–20], and astrocytes [21, 22] have all been reported [15–22]. In contrast to exocrine and endocrine cells,

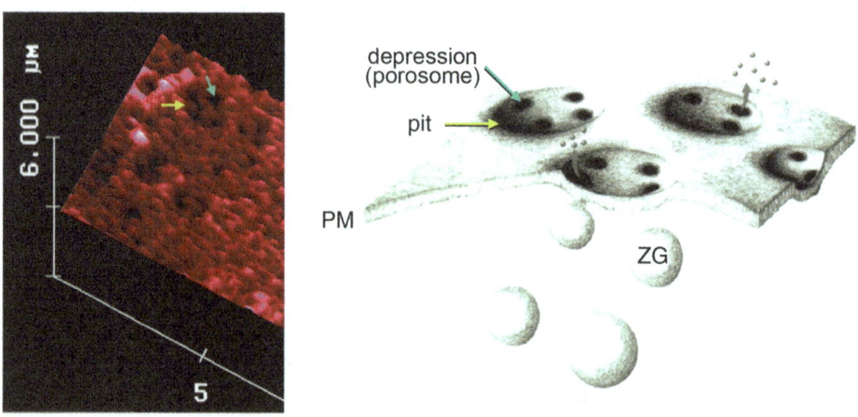

Fig. 1.4 AFM micrograph of *porosomes* at the apical plasma membrane of live pancreatic acinar cells and schematic diagram depicting the structures. To the left is an AFM micrograph of the apical plasma membrane of a live pancreatic acinar cell demonstrating the presence of pits (yellow arrow) with *porosomes* within (blue arrow). To the right is a schematic drawing demonstrating pits and cup-shaped *porosomes* where zymogen granules (ZG), the secretory vesicles in exocrine pancreas dock and transiently fuse to release intra-vesicular contents composed of digestive enzymes from the cell [8]. ©Bhanu Jena

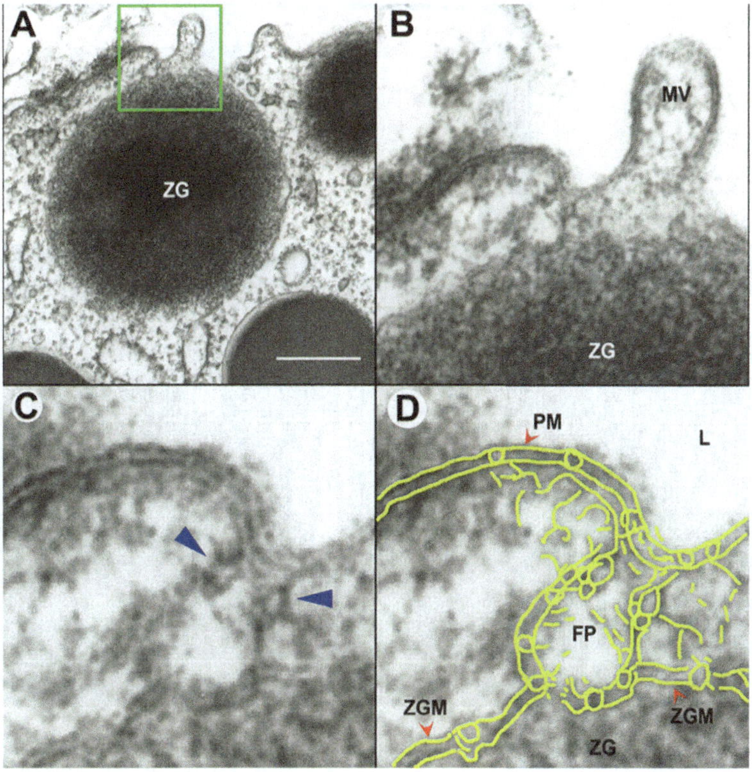

Fig. 1.5 Transmission electron micrograph of a *porosome* at the apical plasma membrane in pancreatic acinar cell associated with a docked and fused secretory vesicle (zymogen granule or ZG). (**a**) Part of the apical end of a pancreatic acinar cell demonstrating within the green square the presence of a porosome and a ZG fused at its base (*Bar = 400 nm* only in Fig. **a**). (**b**) The area within the green square in (Fig. **a**) has been enlarged to show the apical microvilli (MV) and a section through the porosome and the ZG. Note the ZG membrane (ZGM) bilayer is fused with the bilayer of the porosome base facing the cell cytosol. (**c**) At higher magnification the porosome-associated ZG shows in greater detail the porosome bilayer and cross section through the three protein rings (which appear as knobs in either side of the cup-shaped porosome structure). The thicker ring (blue arrowhead) present close to the opening of the porosome to the outside may regulate dilation of the pore opening. The third ring at the base of the porosome is fused with the ZGM and may represent the t-/v-SNARE rosette or ring complex. (**d**) Yellow outline of the ZG fused porosome complex (FP) demonstrates the continuity of the ZG membrane with the plasma membrane (PM) at the apical end of the pancreatic acinar cell facing the lumen (L). The exact points of contact and fusion of the ZGM with the membrane at the porosome base are clearly seen in the electron micrograph [13]. ©Bhanu Jena

the porosomes reported in neurons and astrocytes are smaller cup-shaped lipoprotein structures measuring just 15 nm and possess a unique central plug that exhibits vertical dynamics for the rapid establishment of the fusion pore and its closure (Figs. 1.7, 1.8, and 1.9).

The porosome structure at the cell plasma membrane, as viewed from the cytosolic compartment of the pancreatic acinar cell [13] (Fig. 1.10), and in neuronal

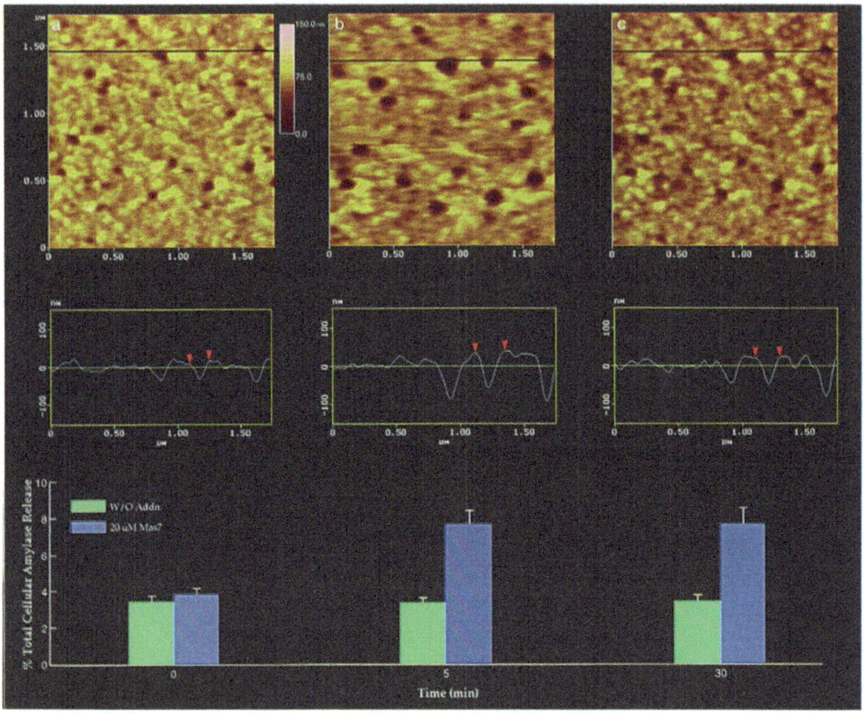

Fig. 1.6 AFM micrograph combined with amylase assays showing porosome dynamics and amylase secretion following stimulation of secretion of pancreatic acinar cells. (*Top Panel*) Several porosomes (red circular structures) at the apical plasma membrane of a pancreatic acinar cell are shown. The scan line across three porosomes in this top panel is represented graphically in the *Center Panel* and defines the diameter and relative depth of each of the porosomes. The porosome in the center is represented by red arrowheads. (*Bottom Panel*) Percent total cellular amylase release in the presence (blue bar) and absence (green bar) of a secretory stimulus (mastoparan 7 (Mas7)). Note the increase in the diameter and relative depth of porosomes following stimulation at 5 min, correlating with an increase in total cellular amylase release. At the 30 min time point post-stimulation, a decrease in diameter and depth of porosomes is demonstrated and is reflected by no further increase in secreted amylase beyond what is observed at the 5-min time point. No significant changes in amylase secretion or porosome diameter were observed in control acini, in either the presence or the absence of the non-stimulatory mastoparan analogue Mas17. The top panel shows AFM micrographs of the time course of porosome dynamics at one representative apical region of pancreatic acinar cell obtained before and after stimulation with Mas7 for up to 30 min [8]. ©Bhanu Jena

synaptosomes (Fig. 1.7f), has also been determined at near nm resolution. Isolated plasma membrane preparations in near physiological buffered solution, when imaged at ultrahigh-resolution using the AFM, reveal scattered circular disks (reverse side of "pits") measuring 0.5–1 μm in diameter, with inverted cup-shaped porosomes within [13]. ZGs ranging in size from 0.4 to 1 μm in diameter are sometimes found docked at the base of one or more of the inverted cup-shaped porosomes. Immuno-AFM studies further confirm that the cup-shaped structures are

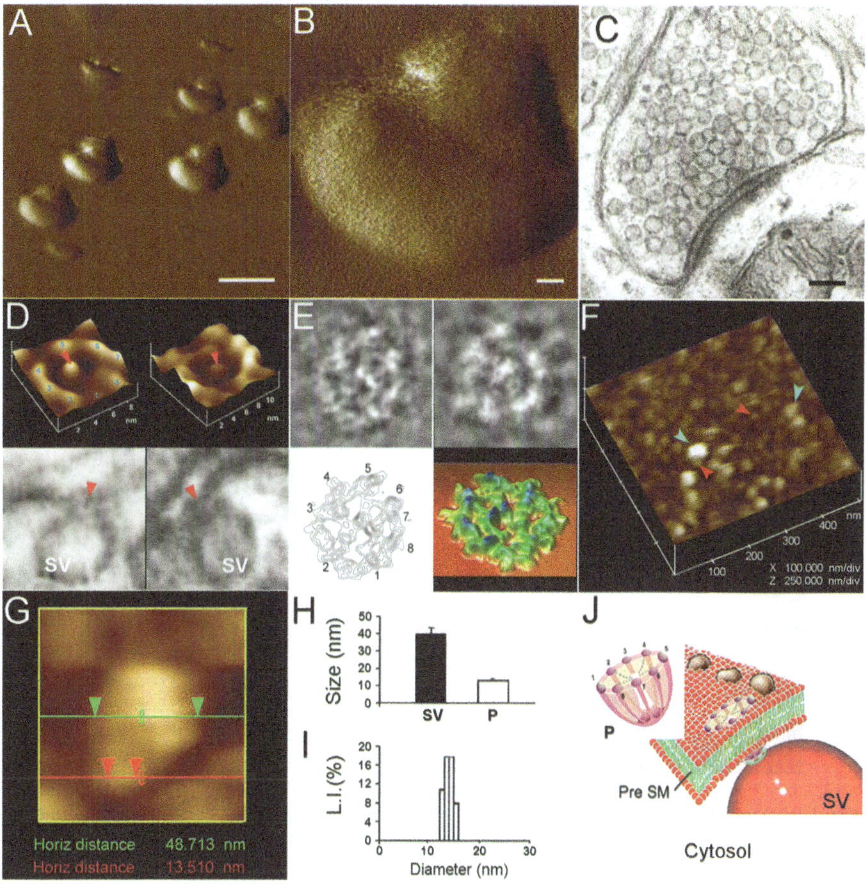

Fig. 1.7 Structure and organization of the neuronal porosome complex at the nerve terminal [16–20]. Low-resolution AFM amplitude image (*Bar* = 1 μm) (**a**) and high-resolution AFM amplitude image (*Bar* = *100* nm) (**b**) of isolated rat brain synaptosomes in buffered solution. (**c**) Electron micrograph of a synaptosome (*Bar* = *100* nm). (**d**) Structure and arrangement of the neuronal porosome complex facing the outside (**d top left**), and the arrangement of the reconstituted complex in PC:PS membrane (**d top right**). Lower panels are two transmission electron micrographs demonstrating synaptic vesicles (SV) docked at the base of cup-shaped porosome, having a central plug (red arrowhead). (**e**) EM electron density and 3D contour maps provide at the nanoscale the structure and assembly of proteins within the neuronal porosome complex. (**f**) AFM micrograph of inside-out membrane preparations of isolated synaptosome. Note the porosomes (red arrowheads) where synaptic vesicles are found docked (blue arrowhead). (**g**) High-resolution AFM micrograph of a synaptic vesicle docked at a porosome in the cytoplasmic compartment of the presynaptic membrane. (**h**) AFM measurements (n = 15) of porosomes (P, 13.05 ± 0.91) and synaptic vesicles (SV, 40.15 ± 3.14) at the cytoplasmic compartment of the presynaptic membrane. (**i**) Photon correlation spectroscopy (PCS) on immunoisolated neuronal porosome complex demonstrates their size to range from 12 to 16 nm. (**j**) Schematic illustration of a neuronal porosome at the presynaptic membrane, showing the eight peripheral ridges connected to the central plug [16–20]. ©Bhanu Jena

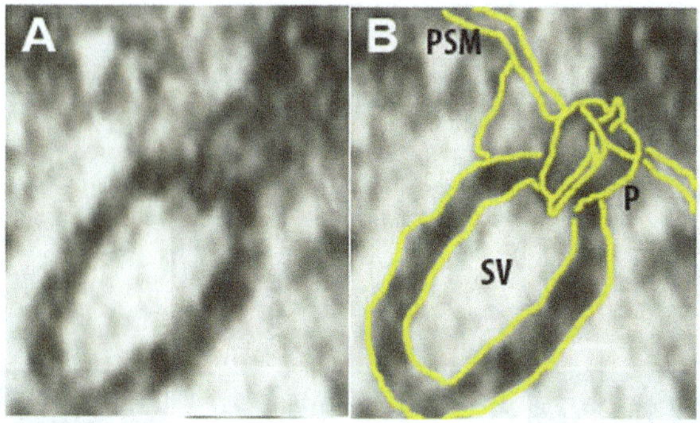

Fig. 1.8 Electron micrograph of a docked synaptic vesicle (SV) at the base of a cup-shaped neuronal porosome complex in rat brain neuron. Neuronal porosome (P) present at the presynaptic membrane (PSM) of a nerve terminal in a rat brain neuron. [*Micron* (2012) 43:948–953. *Courtesy of M. Zhvania*]

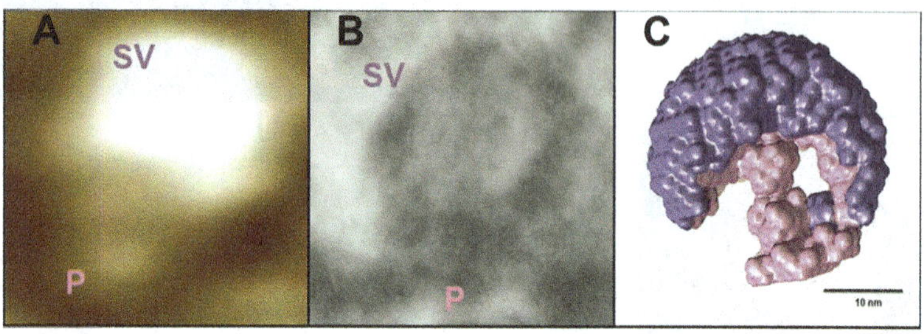

Fig. 1.9 Docked synaptic vesicles at neuronal porosome complex in the presynaptic membrane of the nerve terminal observed using atomic force microscopy (AFM), electron microscopy (EM), and small angle X-ray solution scattering (SAXS). (**a**) AFM micrograph obtained in the fluid of a synaptic vesicle (SV) docked at the cup-shaped porosome complex (P) at the cytosolic compartment of the presynaptic membrane. The 35 nm SV docked to a 15 nm porosome complex is shown. (**b**) An EM micrograph of a 35 nm SV docked to a 15 nm porosome complex (P) at the presynaptic membrane [16]. Note the central plug of the porosome complex in the electron micrograph. (**c**) The averaged SAXS 3-D structure of synaptic vesicle (purple) docked at the cup-shaped neuronal porosome complex (pink) at the presynaptic membrane in isolated synaptosomes is presented [19]. Multimodal imaging using AFM, EM, and SAXS all demonstrate similarity in the docking and interaction of synaptic vesicles at the neuronal porosome complex in the presynaptic membrane. ©Bhanu Jena

indeed secretory portals where secretory vesicles dock and transiently fuse to release intra-vesicular contents. Target membrane proteins, SNAP-23 [23] and syntaxin [24] (t-SNARE) and secretory vesicle-associated membrane protein v-SNARE or VAMP [25], are part of the conserved protein complex involved in the fusion of opposing bilayers in cells [26–29]. The t-SNARE protein SNAP-23 is present in pancreatic acinar cells [21]. A polyclonal monospecific SNAP-23 antibody recognizes a single 23 kDa protein in Western Blots of pancreatic plasma membrane fraction. Immuno-AFM studies demonstrate selective localization of SNAP-23 to the

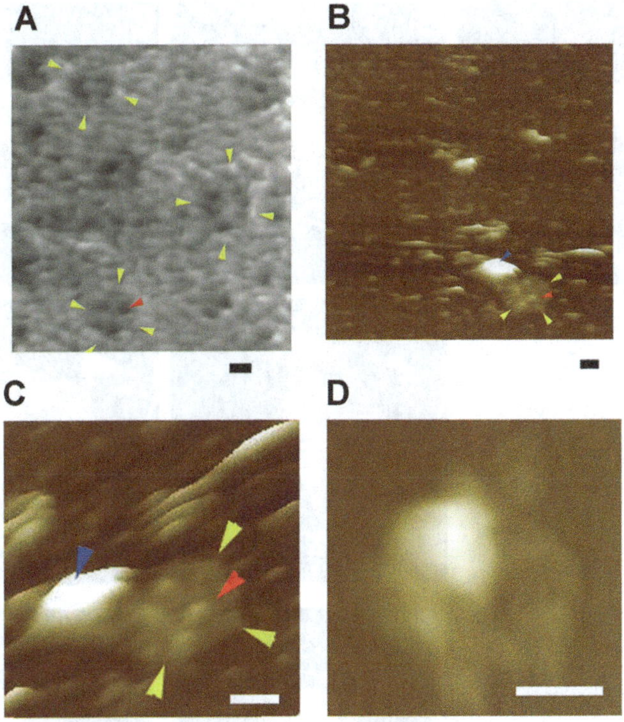

Fig. 1.10 AFM micrographs of porosomes within "pits" at the surface of the apical plasma membrane in live pancreatic acinar cells and at the cytosolic compartment of an isolated pancreatic plasma membrane preparation. (**a**) Several circular "pits" (yellow arrowheads) with porosomes within (red arrowheads) are seen in this AFM micrograph of the apical plasma membrane in a live pancreatic acinar cell. (**b**) AFM micrograph of the cytosolic compartment of an isolated pancreatic plasma membrane preparation depicting a "pit" (yellow arrowheads) containing several inverted cup-shaped porosome structures (red arrowhead) within and associated with a ZG (blue arrowhead). (**c**) The "pit" and inverted porosomes in B are shown at higher magnification. (**d**) AFM micrograph of another "pit" with inverted porosomes within and associated ZG (*Bar = 200 nm*) [13]. ©Bhanu Jena

base of the porosome cup [10] (Fig. 1.11), confirming its function as a secretory portal.

The unequivocal demonstration that porosomes are cellular secretory portals was made using immuno-AFM studies, initially in the exocrine pancreas [9] (Fig. 1.12), followed by studies in GH-secreting somatotrophs of the pig pituitary gland [12]. Localization at porosomes of 30 nm gold-conjugated antibody to the secretory protein amylase provided the first direct evidence that secretion occurs through porosomes. Similarly, in GH-secreting cells of the pituitary gland, 30 nm gold-tagged GH-specific antibody was demonstrated to selectively localize at porosome openings following stimulation of GH secretion [12], establishing porosomes as secretory portals in cells. Porosomes have also been identified and the dynamics of their structure–function examined in insulin-secreting β-cells of the endocrine pancreas.

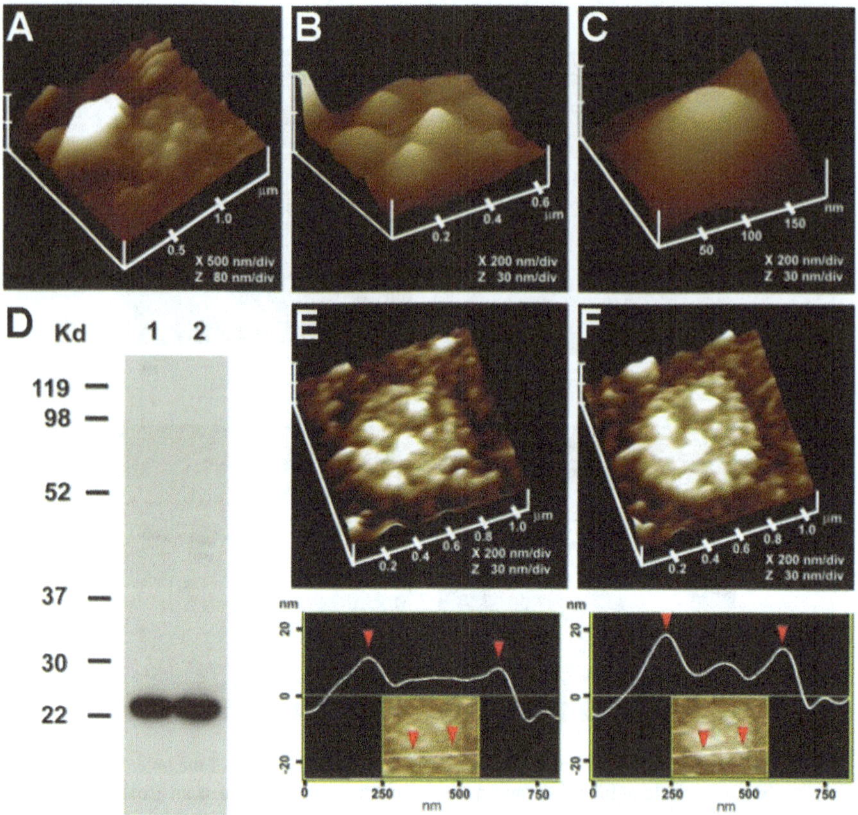

Fig. 1.11 Morphology of the cytosolic compartment of the porosome complex and associated t-SNAREs in the exocrine pancreas revealed using AFM. (**a**) AFM micrograph of isolated plasma membrane preparation reveals the cytosolic compartment of a pit with inverted cup-shaped porosomes. Note the 600 nm in diameter ZG at the left-hand corner of the pit. (**b**) Higher magnification of the same pit demonstrates the presence of 4–5 porosomes within. (**c**) The basolateral end of a single porosome facing the cytosolic compartment of the synaptosome is depicted in the AFM micrograph. (**d**) Immunoblot analysis of 10 μg and 20 μg of pancreatic plasma membrane preparations, using SNAP-23 antibody, demonstrates a specific 23 kDa immunoreactive band. (**e, f**) The cytosolic compartment of the plasma membrane demonstrates the presence of a pit with a number of porosomes within, shown before (**e**) and after (**f**) addition of the SNAP-23-specific antibody. Note the increase in height of the porosome base revealed by section analysis (bottom panel), demonstrating localization of the t-SNARE protein SNAP-23 antibody to the base of the porosome [10]. ©Bhanu Jena

In pancreatic β-cells, porosomes range in size from 100 to 130 nm. Similar to GH and pancreatic acinar cells, exposure of β-cells to elevated levels of glucose results in dilation of the porosome opening to the outside of the cell and increased insulin release. Exposure of β-cells to cytochalasin D results in a decrease in the porosome opening and a loss of insulin secretion.

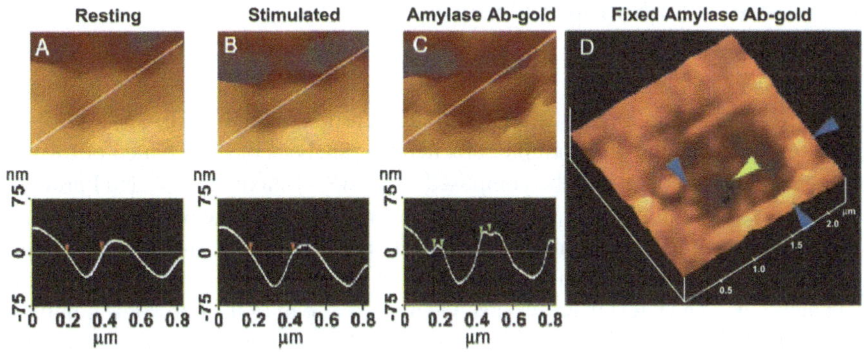

Fig. 1.12 Intra-vesicular amylase is released through the porosome complex in pancreatic acinar cells following stimulation of secretion. (**a**, **b**) AFM micrograph and section analysis of a pit and two of the four porosome openings demonstrating enlargement following stimulation of cell secretion in acinar cell of the exocrine pancreas. (**c**) Exposure of live acinar cells to 30 nm gold-conjugated-amylase antibody (Ab) results in specific localization of the gold particles near the porosome opening to the outside. (**d**) AFM micrograph of a pit with porosome (yellow arrowhead) with immunogold localization (blue arrowhead), demonstrating porosome to be secretory portals [9]. ©Bhanu Jena

The size, shape, and 3D contour map of immunoisolated porosome complex have also been determined using both negative staining EM and AFM studies in the exocrine pancreas [13], neurons [16–18] (Fig. 1.7d, e), and astrocytes [22]. The immunoisolated porosome complex has further been structurally and functionally reconstituted into artificial lipid membrane (Fig. 1.7d). Transmission electron micrographs of pancreatic porosomes reconstituted into liposomes exhibit a 150–200 nm cup-shaped basket-like morphology, similar to its native structure observed in cells and when co-isolated with ZG preparations [13]. Additionally, AFM, EM, and electron density measurements followed by contour mapping and 3D topography [30–33] of the neuronal porosome provide a further understanding of the arrangement of proteins at the nm scale within the porosome (Fig. 1.7e). Photon correlation spectroscopy (PCS) of isolated neuronal porosomes further confirms their 15 nm size (Fig. 1.7i). Results from these studies demonstrate that proteins at the central plug of the neuronal porosome interact with proteins at the periphery of the complex, exhibiting an eightfold symmetry. Furthermore, at the base of the neuronal porosome complex, where synaptic vesicles transiently dock and fuse, possibly SNARE proteins are assembled in a ring or rosette conformation. This possibly neuronal SNARE ring at the porosome base appears to be composed of three SNARE pairs [34, 35] to establish a 1–1.5 nm in diameter channel for the release of neurotransmitters from fused synaptic vesicles. *Details on the establishment of the t-/v-SNARE rosette or ring complex in fusion pore formation and secretory vesicle volume increase for the measured release of vesicular contents during the porosome-mediated "kiss-and-run" mechanism of cell secretion is discussed later in the chapter.*

Porosome: Isolation and Composition

The porosome complex has been isolated using both immuno-pull-down using t-SNARE-specific antibodies and gel filtration chromatography [10, 15–17, 20]. In the exocrine pancreas, for example, biochemical analysis of the isolated porosome demonstrates the complex to be composed of SNAP, syntaxin, cytoskeletal proteins actin, α-fodrin, and vimentin, calcium channels β3 and α1c, the SNARE disassembly ATPase protein *N*-ethylmaleimide sensitive factor (NSF), and the membrane fission GTPase protein dynamin, among others [10, 13, 20]. Similarly, chloride ion channels ClC2 and ClC3 have also been identified as part of the porosome complex, critical to its function. Studies using immunoanalysis and mass spectrometry demonstrate the presence of the cystic fibrosis transmembrane conductance regulator (CFTR) in the porosome complex in human airways epithelia, shedding light on the possible regulatory role of CFTR on the quality of mucus secretion via the porosome complex [3]. The heat shock chaperone protein Hsp90 has been determined to compose the insulin-secreting porosome complex in beta cells and demonstrates attenuation of insulin secretion when inhibited [15]. Isoforms of the various proteins identified within the porosome complex have also been determined using 2D-BAC gels electrophoresis. For example, three isoforms each of the calcium ion channel and vimentin are found in porosomes. Using yeast two-hybrid analysis and immunoisolation studies, the presence and direct interaction between some of these proteins with t-SNAREs within the porosome complex have also been established [36]. Besides proteins, studies report the critical role of lipids such as cholesterol on neuronal porosome assembly [17]. A significant inhibition in the interactions between porosome-associated t-SNAREs and calcium channels is observed following the depletion of membrane cholesterol [17]. Since calcium is critical to SNARE-induced membrane fusion, the loss of interaction between SNAP-25, syntaxin-1, and calcium channels at the neuronal porosome complex could compromise neurotransmission at the nerve terminal.

Among the porosomes studied in different tissues, the chemistry of the neuronal porosome complex is the most well documented, primarily due to its abundance and small size. Determination of the composition of the neuronal porosome proteome using immunoisolation and gel filtration chromatography, followed by tandem mass spectrometry demonstrate 30 proteins to constitute the complex, some presented in Table 1.1 [20]. The association and dissociation of proteins at the porosome following stimulation of cell secretion demonstrates both its stable presence and reflects on the dynamic nature of the nanomachine [20]. Isolated porosomes from rat brain synaptosomes demonstrate the presence of several proteins, among them SNAP-25, the P/Q-type calcium channel, actin, syntaxin-1, synaptotagmin-1, vimentin, the *N*-ethylmaleimide-sensitive factor (NSF), the chloride channel CLC-3, and the alpha subunit of the heterotrimeric GTP-binding $G_{\alpha o}$ protein [16, 17, 20]. Some of the identified proteins have previously been implicated in cell secretion [10, 13, 19, 20, 37, 38]. Several neuronal porosome-associated proteins have also been implicated in neurotransmitter release and various neurological disorders [39–89].

Table 1.1 Proteins identified in purified neuronal porosomes

Gene symbol	MW	Protein name	Found in earlier studies
ACTB	42 kDa	Actin, cytoplasmic 1	x, *
AT1A3	112 kDa	Sodium/potassium-transporting ATPase subunit alpha-3	
AT2B1	139 kDa	Plasma membrane calcium-transporting ATPase 1	
AT2B2	137 kDa	Plasma membrane calcium-transporting ATPase 2	
BASP1	22 kDa	Brain acid soluble protein 1	
CAP1	52 kDa	Adenylyl cyclase-associated protein 1	
CN37	47 kDa	2′,3′-cyclic-nucleotide 3′-phosphodiesterase	
DPYL2	62 kDa	Dihydropyrimidinase-related protein 2	
DPYL3	62 kDa	Dihydropyrimidinase-related protein 3	
DPYL5	62 kDa	Dihydropyrimidinase-related protein 5	
GLNA	42 kDa	Glutamine synthetase	
GNAO	40 kDa	Guanine nucleotide-binding protein G(o) subunit alpha	x, *
NCAM1	95 kDa	Neural cell adhesion molecule 1	
NSF	83 kDa	Vesicle-fusing ATPase	*
RAB3A	25 kDa	Ras-related protein Rab-3A	
RTN3	102 kDa	Reticulon-3	
RTN4	126 kDa	Reticulon-4	
SNP25	25 kDa	Synaptosomal-associated protein 25	X, *
STX1A	33 kDa	Syntaxin-1A	*
STX1B	33 kDa	Syntaxin-1B	*
STXB1	68 kDa	Syntaxin-binding protein 1	
SYN2	63 kDa	Synapsin-2	
SYPH	33 kDa	Synaptophysin	
SYT1	47 kDa	Synaptotagmin-1	*
TBA1A	50 kDa	Tubulin alpha-1A chain	X
VAMP1	13 kDa	Vesicle-associated membrane protein 1	
VAMP2	13 kDa	Vesicle-associated membrane protein 2	
VATB2	57 kDa	V-type proton ATPase subunit B, brain isoform	

Purified rat brain porosomes from two separate experiments were analyzed by LC-MS/MS on both LTQ and QSTAR XL. Only proteins identified in both samples are reported here all of which had protein confidence ≥95% with at least two unique peptides each having 95% confidence or above. Proteins also found in earlier immunoisolation studies are marked in a separate column **x** indicating proteins identified using MALDI-TOF/TOF and ∗ indicating proteins identified using immunoblot analysis (*Taken from Lee* et al., *2012, ref.* 20)

Neuronal porosome-associated proteins like tubulin, myosin 7b, spectrin beta chain, creatine kinase, dystrophin, langerin, GTPase activating protein (GAP), intersectin 1, and myosin heavy chain 1 have been implicated in neurotransmission and neurological disorders. Cross-talk between some of the porosome-associated proteins has been reported. Since mass spectrometry has demonstrated the association of intersectin 1, for example, with the porosome complex, and since intersectin 1 is known to interact with dynamin [21], the association of dynamin with the porosome was hypothesized and later confirmed [20]. Immunoblot analysis of the SNAP-25 immuno-pull-down from solubilized rat brain homogenate demonstrates the presence of dynamin and its increased association with the porosome complex following stimulation of neurotransmitter release [20]. An important aspect of the neuronal porosome complex that emerges from these studies is the dynamic nature of the complex during stimulated neurotransmitter release. Similarly, immunoblot analysis of the SNAP-25 immunoisolate from solubilized rat brain homogenate obtained from control and KCl-stimulated brain slices demonstrates the increased association of dynamin and dissociation of $G_{\alpha o}$ with the neuronal porosome complex following stimulation [20]. It is important to note that plasma membrane calcium-transporting ATPase 1 and 2 are both found in the neuronal porosome complex, suggesting their involvement in ATP-mediated expulsion of the extracellular calcium that had entered the cell during stimulated secretion. Similarly, sodium/potassium-transporting ATPase subunit alpha-3 association with the porosome suggests its role in ATP-coupled exchange of sodium and potassium across the plasma membrane (in opposite directions), creating the required electrochemical gradient for maintaining the membrane resting potential as well as for the regulation of cell volume during the secretory process. Additionally, 2′,3′-cyclic-nucleotide 3′-phosphodiesterase which has been reported to associate with microtubules and has microtubule-associated protein-like activity is part of the neuronal porosome complex. At the porosome, 2′,3′-cyclic-nucleotide 3′-phosphodiesterase can link tubulin to the cell membrane and participate in regulating the distribution of microtubules in the cytoplasm [9]. Hence, the presence of 2′,3′-cyclic-nucleotide 3′-phosphodiesterase at the porosome would be critical to the structural integrity of the nanomachine. Similarly, the dihydropyrimidinase-related proteins present in the porosome complex may play a critical role in cytoskeletal remodeling and cell polarity [11]. Exactly, how these proteins would be involved in neurotransmitter release remains to be established. Similarly, NSF and dynamin at the porosome complex would be required for t-/v-SNARE complex disassembly by NSF and fission of the established continuity between the lipid vesicle membrane and membrane at the porosome base. Since SNAREs are present at the porosome base [13, 34], and calcium channels have been demonstrated to physically interact with SNAREs at the porosome [36], suggests the presence of both proteins at the base of the complex. Similarly, NSF and dynamin would also be present at the porosome base, since they would both be required for the dissociation of the synaptic vesicle from the porosome base following secretion. The function of the central plug in the neuronal porosome complex has been hypothesized to involve motor proteins that help regulate its vertical movement and enable the rapid opening and closing of the fusion pore at the base of the complex. This hypothesis is based on the observation of the central plug at various vertical positions within the porosome complex as observed using AFM.

To further our understanding of the molecular architecture and associations between proteins within the neuronal porosome complex, known interactions between proteins have been carefully examined using the STRING 9.0 database

[90] search (Fig. 1.13) [20]. The STRING maps generated identify two clusters of protein–protein interactions in the neuronal porosome proteome (Fig. 1.13) [20]. In Fig. 1.13, the protein–protein interaction cluster to the left represents primarily cytoskeletal and signaling proteins, whereas those that cluster to the right in the figure represent proteins that are primarily involved in membrane fusion. This is not surprising since the second cluster includes both SNAREs, their associated

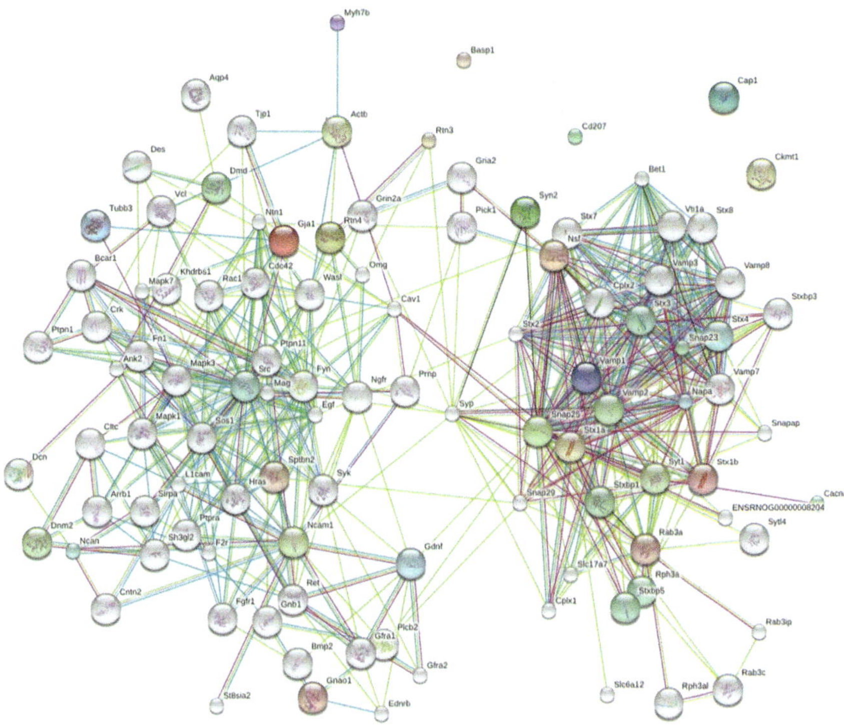

Fig. 1.13 Schematic drawing depicting the evidence view of predicted interactions between identified proteins within the neuronal porosome complex [20]. These interactions are generated from inputs of the identified proteins in the neuronal porosome, using STRING 9.0 (Szklarczyk et al., 2011) [90]. STRING 9.0 is a database of known and predicted protein interactions. The interactions include direct (physical) and indirect (functional) associations derived from genomic, high-throughput, conserved co-expression, and earlier knowledge of known protein–protein interactions. Note the two clusters of protein–protein interactions identified in the neuronal porosome complex. The cluster to the left and most likely present at the apical end of the porosome cup are cytoskeletal structure and signaling proteins. The cluster to the right represents proteins that are primarily involved in membrane fusion including SNARE proteins and calcium channels, and therefore their location would be at the basal part of the porosome cup facing the cytosol. Interestingly, heterotrimeric GTP-binding protein and the GTP-binding membrane fission protein dynamin (Dnm2) are present in the left cluster. The microtubule-associated protein intersectin 1 known to interact with dynamin could enable its translocation to the porosome base to fission the fused synaptic vesicle at the porosome base following completion of neurotransmission. The confidence of the predicted functional interactions shown is >99%. ©Bhanu Jena

regulatory proteins, as well as calcium channels, suggesting their location at the porosome base facing the cytosol. Interestingly, heterotrimeric GTP-binding protein and the GTP-binding membrane fission protein dynamin (Dnm2) are present in the left cluster. The presence of dynamine in the left cluster could be due to the fact that it is a microtubule-associated protein. The involvement of Dmn2 in the fission of the neck of fused vesicles at the porosome base, however, would require its translocation to the porosome base for effective use.

In addition to proteins, our understanding of the composition and distribution of lipids within the neuronal porosome complex has also progressed (Table 1.2) [17, 38, 91]. Earlier studies demonstrate the critical role of cholesterol on the integrity and function of the neuronal porosome complex and cholesterol's influence on the assembly of t-/v-SNARE complex [17, 38, 91]. Results from studies [17] demonstrate a significant inhibition in the interactions between porosome-associated t-SNARE proteins and calcium channels following depletion of membrane cholesterol. Since calcium is critical to SNARE-induced membrane fusion, the loss of such interactions between SNAP-25, syntaxin-1, and calcium channel observed in the absence of cholesterol would compromise neurotransmitter release via

Table 1.2 Major lipid species associated with the neuronal porosome complex identified using mass spectrometry

Lipid Species	Porosome	Synaptosome	AQP1
Cer (ceramides)	**4.82**	0.00	0.00
Cer-P (ceramide phosphates)	**0.48**	0.00	0.00
LPC (cysophosphatidylcholines)	0.00	0.20	0.00
PC (phosphatidylcholines)	4.94	18.50	13.62
SM (sphingomyelins)	0.00	13.78	0.07
LPG (lysophosphatidylglycerols)	0.00	0.50	0.00
LPIP (lysophosphatidylinositol phosphates)	**23.61**	2.81	15.45
LPIP2 (lysophosphatidylinositol diphosphates)	**0.34**	0.14	0.00
LPS (lysophosphatidylserines)	0.00	0.32	0.00
PA (phosphatidic acids)	0.40	2.53	0.00
PE (phosphatidylethanolamines)	0.00	0.94	0.00
PG (phosphatidylglycerols)	0.34	5.19	0.00
PI (phosphatidylinositols)	0.74	6.86	0.00
PIP (phosphatidylinositol phosphates)	1.61	3.23	0.00
PIP2 (phosphatidylinositol diphosphates)	2.98	5.16	0.00
PS (phosphatidylserines)	2.31	20.36	2.21
DAG (diacylglycerols)	**393.74**	2.12	675.89
MADAG (1-alkyl-2-acylglycerols)	**3.00**	0.51	0.00
TAG (triacylglycerols)	0.00	0.19	0.00

Both the porosome and AQP1 (control) were purified from isolated synaptosome preparation. Note the enriched presence of Cer, Cer-P, LPIP, LPIP2, DAG, and MADAG in the isolated neuronal porosome complex. Also note the specific association of PI, PIP, and PIP2 in the neuronal porosome complex and their relative absence in the AQP1 immunoisolates. Protein normalized lipid data is presented as ng/μg protein. (*Taken from Lewis* et al., *2014, ref.* 38) porosomes at the nerve terminal. The lipid composition of isolated neuronal porosome has also been determined using mass spectrometry (Table 1.2) [38]. Using lipid-binding assays, the affinity of porosome-associated syntaxin-1A to various lipids has been determined [38]. Mass spectrometry demonstrates the presence of phosphatidylinositol phosphates (PIPs) and phosphatidic acid (PA) among other lipids, and the enriched presence of ceramide (Cer), lysophosphatidylinositol phosphates (LPIP), and diacylglycerol (DAG) associated with the neuronal porosome complex [38]. Lipid-binding assays further demonstrate the association of cardiolipin with neuronal porosome and confirm the association of PIPs and PA with the complex. The ability of exogenous PA to alter protein–protein interaction and neurotransmitter release is further demonstrated from the study [38].

The presence and requirement of PIP_2 at the synaptic vesicle docking site of the presynaptic membrane have also been reported [92]. In recent years, at least 11 lipid-binding domains have been identified in proteins, which include C1, C2, PH, FYVE, PX, ENTH, ANTH, BAR, FERM, PDZ, and tubby domains [38]. The lipid-binding C2 domain is present in the porosome-associated protein synaptotagmin. The 3D structure of the C2 domain of synaptotagmin has been reported to form an eight-stranded beta-sandwich constructed around a conserved 4-stranded motif designated as a C2 key, and calcium binds to the cup-shaped depression formed by the N- and C-terminal loops of the C2-key motif. Structural analyses of several C2 domains have been shown to consist of similar ternary structures in which three Ca^{2+}-binding loops are located at the end of an 8-stranded antiparallel beta-sandwich. Another porosome protein spectrin beta represents an important group of actin, dystrophin, and lipid-binding proteins involved in membrane anchoring. Similarly, dystrophin, the cytoskeletal porosome protein that confers resistance to contraction–relaxation cycles by interacting with other cytoskeletal and membrane proteins, possesses 24 spectrin-like repeats and phospholipid binding domains. The effect of membrane lipids on ion channel functions, including calcium channels, has also been reported. Studies show that the activity of calcium channels could be modulated by lipid domain formation upon minor changes in the composition of membrane lipids. Recent studies involving the crystal structure of a lipid-G-protein-coupled receptor demonstrate that the lysophospholipid sphingosine 1-phosphate modulates receptor activity. In this way, the structure–function of several neuronal porosome proteins could either be directly or indirectly modulated by a variety of membrane lipids. Keeping in mind that all membrane proteins are known to be solvated by a shell of lipid molecules referred to as annular lipids that interact with the membrane-penetrating hydrophobic domain of the protein, these annular lipids likely contribute to protein phase separation within the porosome complex. Such phase separation transitions and the yet to be identified lipid and protein switches involved would provide the enthalpy for the molecular assembly of the porosome complex and the observed association and dissociation of different proteins within the complex during the secretory process. Similarly, lipid molecules that are also found bound between transmembrane α-helices of membrane proteins within the bilayer are referred to as non-annular lipids. A number of studies investigating the crystal structures of membrane proteins report the presence of non-annular bound lipids and cholesterol molecules between transmembrane domains of membrane proteins. Annular lipid binding is dependent on the fatty acyl chain lengths, suggesting that hydrophobic matching between membrane proteins and the surrounding lipid bilayer exerts physical distortions on the α-helical bundle of membrane proteins. It has also been demonstrated that annular lipids readily exchange with the bulk bilayer lipids. In contrast, the mobility of non-annular lipids is somewhat restricted; hence, the precise interaction between proteins and lipids profoundly influences each other's structure–function. For example, studies report that Na-K-ATPase also present in the neuronal porosome complex is highly dependent on acid phospholipids, especially phosphatidylserine (PS) for its stability. The PS-binding site of Na-K-ATPase has been identified to be located between the 8, 9, and 10 transmembrane segments of the α-subunit of the protein, and studies further demonstrate the binding and activation of Na-K-ATPase by neutral phospholipids such as phosphatidylcholine (PC) or phosphatidylethanolamine (PE). Similar to Na-K-ATPase, another neuronal porosome protein Ca-ATPase has also been demonstrated to possess specific sites that bind phosphatidylethanolamine (PE) [38]. A significant contribution to cellular and subcellular structure is known to be imparted by those proteins that cross-link actin, a neuronal porosome protein, or connect actin filaments to cellular membranes. The spectrin superfamily of cytoskeletal proteins including spectrin β and dystrophin, also known neuronal porosome proteins, possesses pleckstrin homology (PH) domains that bind lipids and help serve as membrane anchors. PH repeats are each composed of up to nearly 120 amino acids, and to their basic residues, phosphates of phosphoinositides such as PI-(4,5)-P2, PI-(3,4,5)-P3, PI-(3,5)-P2, and Ins(1,4,5)P3 are known to bind [38]. Observation of the presence of PIP2 in the porosome complex using mass spectrometry and in lipid binding experiments [38] supports such protein–lipid interactions within the neuronal porosome complex [38, 92]. Similarly, the porosome-associated proteins GAP, dynamin, and intersectin 1 possess BAR (Bin/Amphiphysin/Rvs) domains, which are known to bind lipid membranes having high curvature. A typical BAR dimer is composed of a banana-shaped group of six helices having positive charges at the tips and along its concave surface that may mediate phospholipid binding. While BAR domains bind curved phospholipid membrane segments, it is proposed that a subset of BAR domains could physically assist in inducing membrane curvature. Therefore, it is of interest to note that most of the inositol lipid-binding protein domains also bind specific proteins; hence, protein–lipid–protein interactions could influence each other at the porosome complex. A careful comparison of the structural features of lipid-binding domains of membrane proteins demonstrates a high degree of conservation, reflecting on their evolutionary link and interactions, critical to biological function. The important observation and advances made in the past 40 years regarding PI-(1,4,5)P3 as cellular second messengers have further shed light on their involvement in diverse cellular processes, including perhaps the structure and function of the neuronal porosome complex [38].

Porosome: Functional Reconstitution in Lipid Membrane and Live Cells

To test whether the isolated porosome preparations are functional, purified porosomes obtained from acinar cells of the exocrine pancreas [13], from beta cells of the endocrine pancreas [16] and from brain neurons [14], have been structurally and functionally reconstituted into lipid membrane and in live cells [16]. For the functional reconstitution of porosomes into artificial lipid membranes, an electrophysiological bilayer setup (EPC9) having a 200 μm in diameter bilayer membrane separating a cis and a trans compartment (Fig. 1.14) has been utilized [13, 14]. Lipid bilayers are prepared using brain phosphatidylethanolamine (PE) and phosphatidylcholine (PC) obtained from Avanti Lipids (Alabaster, AL). A suspension of PE:PC in 7:3 ratio is used to establish a stable bilayer in a 200 μm in diameter hole in the bilayer preparation cup. Bilayer membranes are established while holding at 0 mV, and once a bilayer is formed and demonstrated to be in the capacitance limits for a

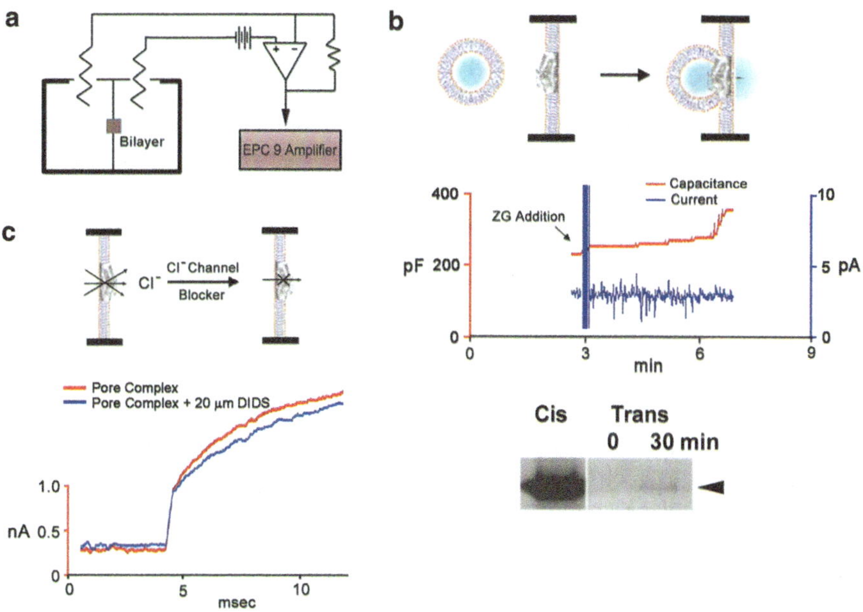

Fig. 1.14 Isolated porosome complexes are functional when reconstituted into artificial lipid bilayer membrane. (**a**) Schematic drawing of the bilayer setup for electrophysiological measurements. (**b**) Zymogen granules (ZGs) added to the *cis* compartment of the bilayer fuse with the reconstituted porosomes, as demonstrated by an increase in capacitance and current activities, and a concomitant time dependent release of amylase (a major ZG content) to the *trans* compartment of the membrane. The movement of amylase from the *cis* to the *trans* compartment of the chamber was determined by immunoblot analysis of the contents in the *cis* and the *trans* chamber over time. (**c**) Electrical measurements in the presence and absence of the chloride ion channel blocker DIDS demonstrate the functional presence of chloride ion channels within the porosome complex and its critical role in porosome-mediated secretion [13]. ©Bhanu Jena

stable bilayer membrane, the voltage is switched to −60 mV. A baseline current is established prior to the addition of isolated secretory vesicles to the cis compartment of the bilayer chamber. An established stable bilayer exhibits a capacitance between 100 and 250 pF. Isolated porosomes are then reconstituted into the bilayer membrane by brushing in isolated porosome preparations (Fig. 1.14). Time-dependent changes in current and capacitance are recorded before and following the addition of secretory vesicles (ZG or synaptic vesicles) to the *cis* compartment of the bilayer chamber. Capacitance and conductance recordings are made using pulse software, EPC9 amplifier, and detection probe from HEKA (Lambrecht, Germany). Aliquots from the cis and trans chamber are collected at different time intervals following the addition of secretory vesicles to the *cis* compartment, to immunochemically (Western Blots) determine the transport of intra-vesicular contents via the porosome from *cis* to the *trans* chamber of the EPC9 apparatus. Electrical activity of the porosome-reconstituted membrane as well as the transport of vesicular contents from the *cis* to the *trans* compartments of the bilayer chambers when monitored demonstrates that the lipid membrane-reconstituted porosomes are indeed functional (Fig. 1.14) [13, 14]. In the presence of calcium, isolated secretory vesicles dock and transiently fuse at the porosome to transfer intra-vesicular contents from the *cis* to the *trans* compartment of the bilayer chamber (Fig. 1.14). A time-dependent transport of the ZG enzyme amylase from *cis* to the *trans* compartment of the bilayer chamber is demonstrated using immunoblot analysis of the buffer in the two compartments. In pancreas, chloride channel activity present at the reconstituted porosome complex is found to be critical to its activity, since the chloride channel blocker DIDS inhibits reconstituted porosome-mediated vesicular content release (Fig. 1.14).

Min6 mouse insulinoma cells exhibit glucose-stimulated insulin secretion. Isolated Min6 porosomes reconstituted into live Min6 cells exhibit an increase in potency and efficacy of glucose-stimulated insulin release within 1 h following reconstitution [16]. This glucose-stimulated insulin release is sustained in these reconstituted cells beyond a 48 h period, demonstrating for the first time, the functional reconstitution of porosomes into live cells [16]. Functional reconstitution of the porosome in live cells holds great promise for future therapeutic applications. For example, cellular secretory defects resulting from impaired porosome functions could be overcome by reconstituting isolated porosomes from healthy cadaveric tissue. The reconstituted nanoscale porosome complex is unlikely to elicit an immune response and therefore in certain situations would be an advantage over tissue transplant.

Establishment of the t-/v-SNARE Rosette or Ring Complex at the Porosome Base

In 1988 Richard Scheller discovered a secretory vesicle-associated membrane protein called VAMP-1 or v-SNARE [25], and then in 1992 he and his team discovered another protein present in the cell plasma membrane called syntaxin [24]. Syntaxin is one of the two target SNARE or t-SNARE proteins. In 1989 Michael Wilson discovered SNAP-25, the other t-SNARE protein [23]. Understanding the properties of the three SNARE proteins in membrane fusion requires a molecular understanding of their interactions when associated with their respective membranes: v-SNARE or VAMP-1 in secretory vesicle membrane and t-SNAREs syntaxin and SNAP-25 at the cell plasma membrane. Since SNAREs are membrane-associated proteins, crystals of membrane-associated SNARE complex are required for X-ray crystallography, which has not been possible due to solubility problems. To circumvent issues associated with the solubility of membrane-associated SNAREs, Axel Brunger and Reinhard Jahn in 1998 truncated the hydrophobic membrane anchoring domains of syntaxin and VAMP to obtain crystals of a non-membrane-associated t-/v-SNARE complex. Utilizing X-ray crystallography, Brunger and Jahn determined the atomic structure of the soluble SNARE complex at 2.4 Å, which they reported in the journal *Nature* [93]. It was unclear, however, whether the structure of the resolved soluble SNARE complex was identical to the native membrane-associated SNARE complex.

To address this issue, we carried out high-resolution structural AFM studies combined with functional electrophysiological measurements. In a study reported in the Biophysical Journal in 2002 [94], our group demonstrated that in the absence of membrane association, SNAREs fail to appropriately bind to each other or establish continuity between the opposing bilayers in the presence of calcium. We demonstrated that VAMP-1 proteins present in one membrane interact with syntaxin and SNAP-25 proteins present in an opposing membrane, to assemble in a rosette or ring configuration, establishing continuity between the opposing bilayers in the presence of calcium (Fig. 1.15). While it had been hypothesized that the interaction between t-SNAREs and v-SNARE present in opposing bilayers may form such rosette or ring structures [99], the experimental evidence and confirmation of this hypothesis were first reported by our group in the 2002 Biophysical Journal paper [94] and further established using high-resolution AFM and EM studies [95–98]. This SNARE rosette arrangement between opposing bilayers during membrane fusion is now widely accepted to be required for membrane fusion during cell secretion [100, 101].

The 15 nm in diameter neuronal porosome complex has a mere 4 nm in diameter base where 30–50 nm in diameter synaptic vesicle transiently dock and fuse to release neurotransmitters. This allows room for only one or two SNARE ring complexes at the neuronal porosome base, each ring composed of merely three SNARE pairs (Fig. 1.16 inset), and a SNARE ring complex channel measuring just 1–1.5 nm in diameter. Unlike exocrine or neuroendocrine secretions, the 1–1.5 nm in diameter

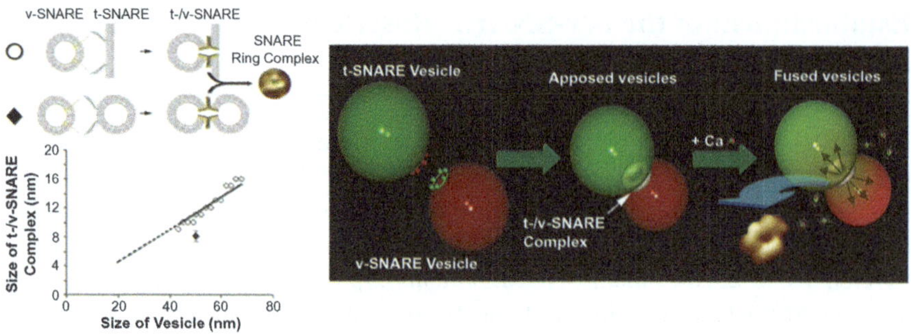

Fig. 1.15 Size of the t-/v-SNAREs ring complex is directly proportional to vesicle size [94–98]. (*Left*) Interaction between t-SNARE-associated membrane with v-SNARE-associated vesicles of different sizes (○) demonstrates the SNARE ring size to be directly proportional to the vesicle size. Hence membrane curvature dictates the size of the t-/v-SNARE ring complex. When a 50 nm in diameter v-SNARE-reconstituted vesicle interacts with a t-SNARE-reconstituted membrane, an 11 nm in diameter t-/v-SNARE ring complex is formed. However, when a 50 nm in diameter v-SNARE-reconstituted vesicle interacts with another 50 nm in diameter t-SNARE-reconstituted vesicle, an 8 nm in diameter t-/v-SNARE ring complex is established (◆), demonstrating membrane curvature to be critical in dictating the size of the t–/v-SNARE ring or rosette **complex**. (*Right*) Schematic drawing depicting the interaction between t-SNAREs and v-SNAREs in opposing vesicles to establish the t-/v-SNARE ring and the establishment of continuity (fusion pore) between the opposing bilayers in the presence of calcium. An AFM micrograph of an actual t-/v--SNARE ring complex is shown as a yellow pseudo-color ring complex to the far right. ©Bhanu Jena

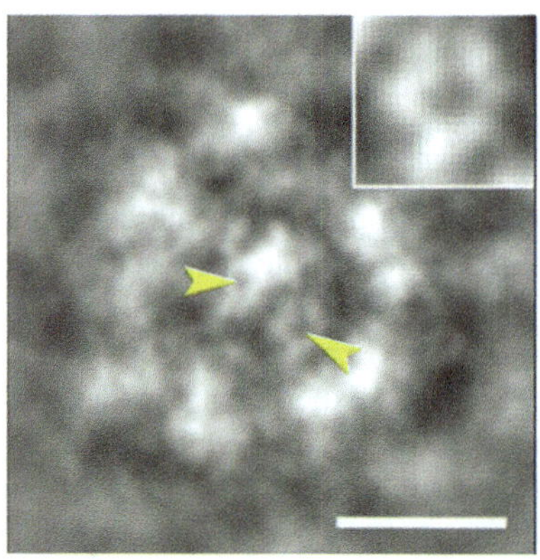

Fig. 1.16 Electron micrograph obtained using transmission electron microscopy performed on negatively stained isolated neuronal porosome complexes. The micrograph demonstrates the presence of two ring complexes at the center (porosome base), each possibly corresponding to three t-/v-SNARE pairs. Note the 15 nm complexes exhibiting a circular profile and having a central plug. Approximately 8 interconnected protein densities are observed at the rim of the structure, which are connected to a central plug via spoke-like tethering elements. At the center of the structure corresponding to the porosome base, there are protein complexes arranged in ring or rosette conformation, likely corresponding to t-/v-SNARE complex established by three SNARE pairs (yellow arrowheads and inset) *Bar = 5 nm* [18]

central channel would be adequate for neurotransmitter release as a result of a build-up of intra-vesicular pressure following synaptic vesicle swelling during neurotransmission.

Vesicle Volume Regulates Porosome-Mediated Fractional Content Release

In the early 1990s it was reported that secretory vesicles undergo an increase in volume during cell secretion [102, 103]. The molecular mechanism underlying the regulation of volume increase in secretory vesicles and the role of this volume increase on secretory vesicle function during cell secretion, however, were poorly understood. Our studies demonstrated that water channels or aquaporins in conjunction with several ion channels present at the secretory vesicle membrane regulate vesicle volume through a GTP-binding G-protein-mediated signaling pathway (Figs. 1.17 and 1.18) [104–110]. The role of various ion channels at the secretory vesicle membrane was also demonstrated using single vesicle patch clamp studies [111]. In 2004 we reported that secretory vesicle volume increase is a requirement for the regulated release of vesicular contents from cells during secretion (Fig. 1.19) [105]. The relative increase in vesicle volume during cell secretion is proportional to the fraction of the intra-vesicular contents released.

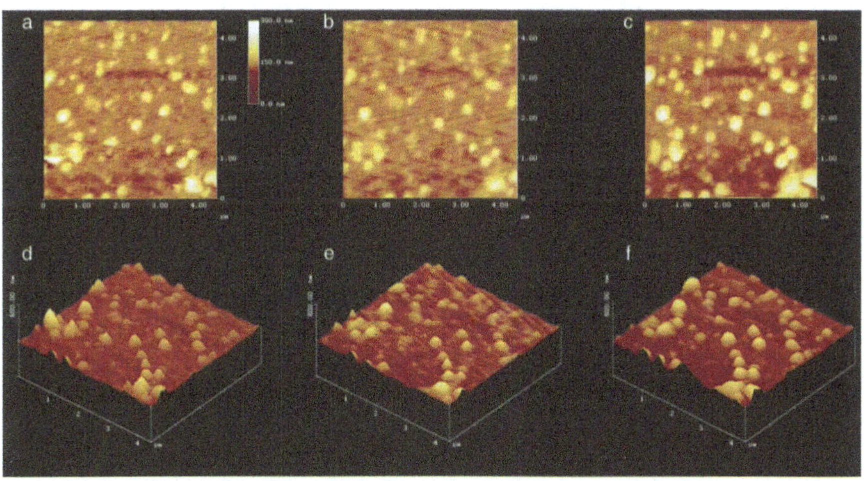

Fig. 1.17 AFM micrograph of time-dependent increase in the size of isolated ZGs in physiological buffered solution following GTP exposure. (**a–c**) 2D AFM images of the same ZGs following exposure to 20 μM GTP at time 0 (**a**), 5 min (**b**), and 10 min (**c**). (**d–f**). The same granules are shown in 3D: the three-dimensional AFM images of the ZGs at time 0, 5 min, and 10 min, respectively, following exposure to GTP [104]. ©Bhanu Jena

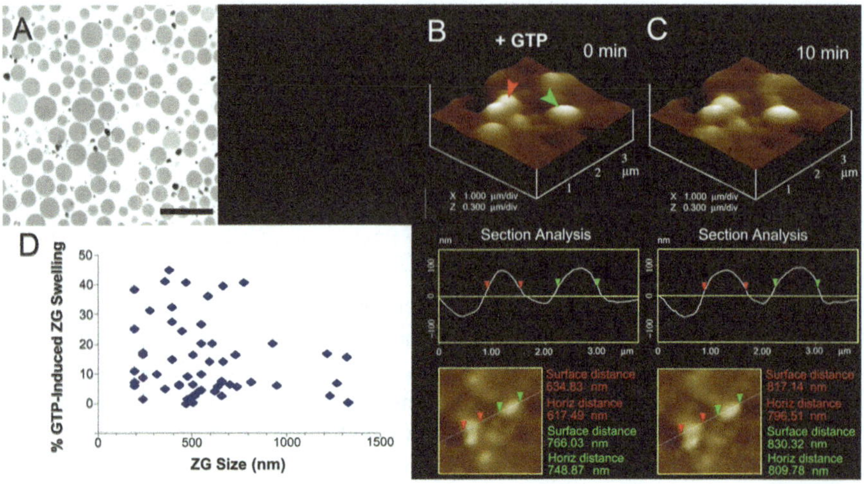

Fig. 1.18 Differential swelling of isolated ZGs is observed following GTP exposure. (**a**) Electron micrograph of isolated ZGs demonstrating purity of the preparation. The black electron dense particles represent inert Percoll used in the preparation of the gradient for ZG isolation. *Bar = 2.5 μm*. (**b**, **c**) Isolated ZGs, on exposure to 20 μM GTP, swell rapidly. As an example, the enlargement of ZGs as determined by AFM section analysis of two vesicles (red and green arrowheads) is presented. (**d**) Varying percent ZG volume increase in response to 20 μM GTP is demonstrated [105]. ©Bhanu Jena

Isolated secretory vesicles, single vesicle patch, and reconstituted swelling-competent proteoliposomes have been utilized [104–111] to determine the mechanism and regulation of secretory vesicle volume. Isolated ZGs from the exocrine pancreas swell rapidly in response to GTP, suggesting rapid water gating into ZGs. Results from studies demonstrate the presence of the water channel aquaporin-1 (AQP1) at the ZG membrane [106] and aquaporin-6 (AQP6) at the synaptic vesicle membrane [107] and their participation in GTP-mediated water entry and vesicle swelling. Further, the molecular regulation of AQP1 at the ZG membrane has been studied [108], providing a general mechanism of secretory vesicle swelling. Detergent-solubilized ZGs are immunoisolated using monoclonal AQP-1 antibody,

Fig. 1.19 (continued) induced, resulting in the expulsion of intra-vesicular contents such as α-amylase to the trans bilayers chamber. (**b**) Capacitance traces of the lipid bilayer from three separate experiments following reconstitution of porosomes (green arrowhead), addition of ZGs to the cis compartment (blue arrowhead), and the red arrowhead represents the 5 min time point after ZG addition. Note the small increase in membrane capacitance following porosome reconstitution, and a greater increase when ZGs fuse at the bilayer membrane. (**c**) In a separate experiment, 15 min after addition of ZGs to the cis compartment, 20 μM GTP was introduced. Note the increase in capacitance demonstrating potentiation of ZG fusion. Flickers in current trace represent activity. (**d**) Immunoblot analysis of α-amylase in the trans chamber fluid at different times following exposure to ZGs and GTP. Note the undetectable levels of α-amylase even up to 15 min following ZG fusion at the bilayer. However, following exposure to GTP, significant amounts of α-amylase from within ZGs were expelled into the trans bilayers chamber [105]. ©Bhanu Jena

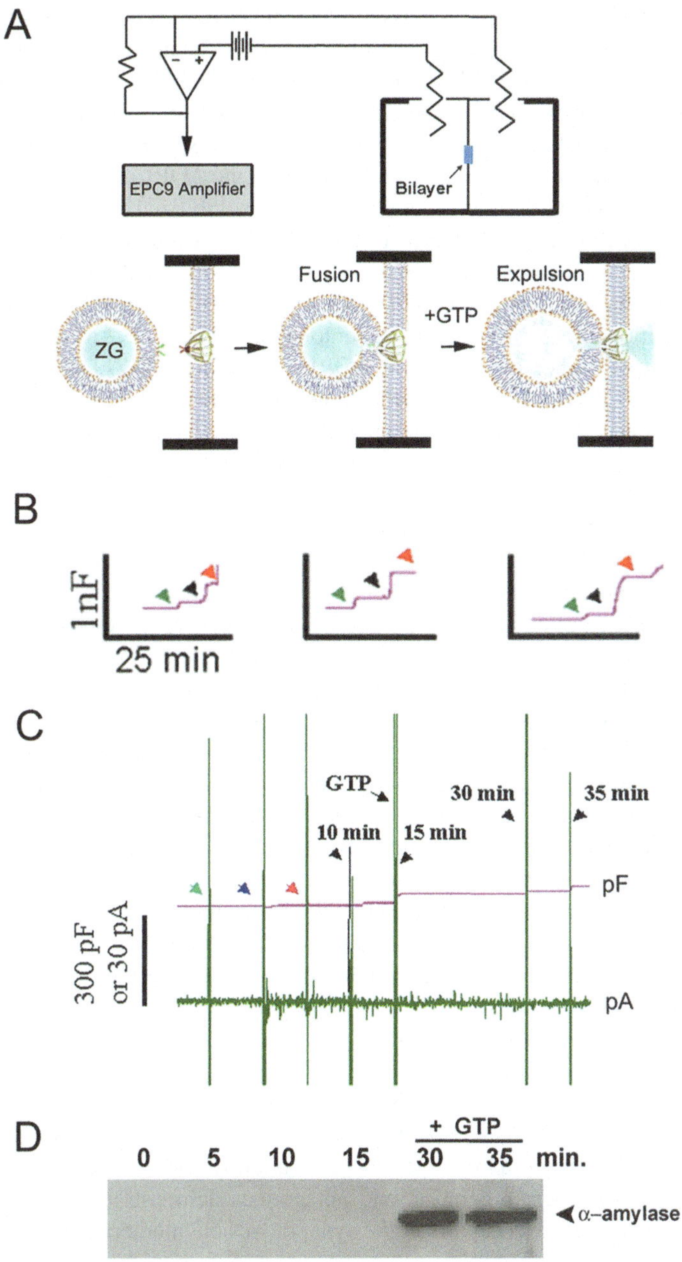

Fig. 1.19 Secretory vesicle swelling is required for porosome-mediated secretion. Fusion of isolated ZGs at porosome-reconstituted bilayer and GTP-induced expulsion of α-amylase is demonstrated. (**a**) Schematic diagram of the EPC9 bilayer apparatus showing the cis and trans compartment of the bilayer chamber. Isolated ZGs when added to the cis compartment fuse at the bilayers-reconstituted porosome. Only following addition of GTP to the cis chamber ZG swelling is

co-isolates AQP1, PLA2, $G_{\alpha i3}$, potassium channel IRK-8, and the chloride channel ClC-2 [108]. Exposure of ZGs to either the potassium channel blocker glyburide or the PLA2 inhibitor ONO-RS-082 blocks GTP-induced ZG swelling. Red blood cells known to possess AQP1 at the plasma membrane also swell on exposure to the $G_{\alpha I}$ agonist mastoparan and respond similarly to ONO-RS-082 and glyburide, as do ZGs [108]. Artificial liposomes reconstituted with the AQP1 immunoisolated complex (vesicle volume regulation complex) from solubilized ZG preparation also swell in response to GTP. Glyburide or ONO-RS-082 is found to abrogate the GTP effect in reconstituted proteoliposomes. AQP1 immunoisolated-reconstituted planar lipid membrane demonstrates conductance, which is sensitive to glyburide and an AQP1 specific antibody. These results demonstrate a $G_{\alpha i3}$-PLA2 mediated pathway and potassium channel involvement in AQP1 regulation at the ZG membrane [108], contributing to ZG swelling. Similarly, AQP-6 involvement has been demonstrated in GTP-induced and G_o-mediated synaptic vesicle swelling in neurons [107]. In addition to elucidating the mechanism of secretory vesicle volume regulation, results from these studies provide the molecular mechanism involved in the regulation of the bidirectional aquaporin water channels.

To characterize ion channels present at the secretory vesicle membrane, we carried out studies using single ZG patch clamp [111]. These studies confirm earlier findings of the presence of both potassium and chloride ion channels at the ZG membrane. In these studies, the electrical activity at the ZG membrane displays a range of sensitivity both to chloride and potassium channel blockers. Whole vesicle conductance is decreased with the addition of the chloride channel blocker DIDS and the ATP K+ channel blocker glyburide, in both vesicle patches and indirect analysis, supporting the hypothesis for the presence of multiple isoforms of a channel [111]. This finding was further confirmed immunochemically using Western Blot analysis and as speculated, the presence of 2 chloride channels, CLC-2 and CLC-3, was demonstrated from these studies [111]. Consistent with pharmacological evidence, the presence of ATP sensitive potassium channel, Kir6.1 was confirmed from Western Blot analysis of isolated ZG preparations.

Since mastoparan, an amphiphilic tetradecapeptide from wasp venom activates G_o protein GTPase and stimulates synaptic vesicle swelling, the presence of β-adrenergic receptor at the synaptic vesicle membrane was hypothesized. Stimulation of G proteins is believed to occur via insertion of mastoparan into the phospholipid membrane to form a highly structured α-helix that resembles the intracellular loops of G protein-coupled adrenergic receptors. Immunoblot analysis of synaptic vesicle using β-adrenergic receptor antibody and vesicle swelling experiments using β-adrenergic agonists and antagonists demonstrates the presence of functional β-adrenergic receptors at the synaptic vesicle membrane [110].

Porosome in Health and Disease

In this section, the neuronal porosome complex and associated proteins are discussed. Cytoskeletal proteins such as actin and the alpha chain of tubulin are two of the several neuronal porosome proteins detected using mass spectrometry. It has long been established that actin is involved in neurotransmission [39]. Similar to the exocrine pancreas demonstrating the collapse of the porosome opening and loss of amylase secretion following exposure to the actin depolymerizing cytochalasin D [8], studies performed on neurons in the presence of latrunculin A, also an actin depolymerizing agent, partially blocked neurotransmitter release at the presynaptic terminal of motor neurons [40]. Actin b protein, a post-translational product of actb mRNA is important in the formation of excitatory synapses, which is promoted by interaction of actb mRNA with Src associated in mitosis (Sam68 protein). Loss in Sam68 diminishes its interaction with actb mRNA leading to lower levels of synaptic actin b, which in turn leads to neurological disorders involved with abnormal synaptic transmissions [41]. Although tubulin involvement in neurotransmission is not fully understood, its association with several synaptosomal proteins at the presynaptic membrane [42, 43] suggests an important role in neurotransmission. Besides actin and tubulin, other important class of proteins identified in the neuronal porosome are the membrane integrated ion-channel proteins that are important in maintaining cellular homeostasis. A good example is the alpha subunit 3 of Na^+/K^+ATPase present in the porosome and plays a critical role in neurotransmission. Na^+/K^+ATPase activity is known to be blocked by dihydroouabain (DHO) [44]. Activation of Na^+/K^+ATPase inhibits synaptic transmission, which results from the secretion of a presynaptic protein "follistatin-like 1" which activates the alpha 1 subunit of the ATPase [45]. Transient blocking of Na^+/K^+ATPase pump using DHO results in an increase in both the amplitude and number of action potentials at the nerve terminal [46]. Interestingly, Na^+/K^+ATPase inhibition is calcium dependent and it has been demonstrated that increased intracellular calcium levels inhibit Na^+/K^+ATPase, which increases excitability of neurons [47]. Similarly, the porosome protein plasma membrane calcium ATPases (PMCA) are an important class of proteins known to be involved in maintaining calcium homeostasis within cells including neurons [48]. PMCA2 has been shown to co-localize in synaptosomes with synaptohysin [48–51]. At the presynaptic membrane, syntaxin-1 also a porosome protein has been demonstrated to co-localize with PMCA2 and the glycine transporter 2 (GlyT2) that is found coupled to the Na^+/K^+ pump, suggesting the presence of a protein complex involved in neurotransmission [49–51]. Similarly, deletion of PMCA2 generates an ataxic phenotype in mice, where the neurons possess prolonged hyperpolarized membrane, resulting from an increase in the basal levels of calcium within these neurons [52]. Mutation in the PMCA2 encoding gene is known to result in homozygous deafwaddler mice (dfw/dfw) and they show high levels of calcium accumulation within their synaptic terminals [53]. NAP-22, also known as BASP-1, is a protein found in the neuronal porosome complex, and due to its localization at the pre- and post-synaptic membranes and also in synaptic vesicles, it has been long speculated to be involved in synaptic transmission [54–56]. NAP-22 is known to bind to the inner leaflet of lipid rafts suggesting interaction with cholesterol. Adenylyl cyclase-associated protein-1 (CAP-1) is known to regulate actin polymerization [57] and both actin and CAP-1 are present in the porosomal complex. In cells depleted of CAP-1 using RNAi, result in lamellipodia growth and F-actin accumulation along with other cytoskeletal abnormalities [58]. In Alzheimer's, the levels of CNPase (2,3-cyclic nucleotide phosphodiesterase) and the heat shock protein 70 (HSP70) are found to increase, while the levels of dihydropyrimidinase-related protein-2 (DRP-2) decrease [59]. Decreased levels of CNPase have been detected in the frontal and temporal cortex of patients with Alzheimer's disease and or Down's syndrome [60, 61]. Low CNPase levels have also been detected in post-mortem anterior frontal cortex in schizophrenic patients [62]. Additionally, an allele that is associated with low levels of CNPase is also reported to be linked to schizophrenia [63]. CNPase is also detected as a marker for oligodendroglia and myelin and several diseases associated with low levels of CNPase indicating low myelin on neurons [64]. CNPase positive cells have been shown to increase in the corpus callosum of rats exposed to an enriched environment, meaning given to perform a task [65]. Similarly, alterations in the levels of several of the SNARE proteins are associated with various neurological disorders. For instance, SNAP-25 and synaptophysin are significantly reduced in neurons of patients with Alzheimer's disease [66–68]. Mice that are SNAP-25 (+/−) show disabled learning and memory and exhibit epileptic like seizures [69]. Overexpression of SNAP-25 also results in defects in cognitive function [70]. Loss of SNAP-25 is also associated with Huntington's disease and a reduction in rabphilin 3a, another protein involved in vesicle docking and fusion at the presynaptic membrane [71]. SNARE knockout mice are neonatally lethal and mice with a dominant mutation in SNAREs are known to develop ataxia and show impairment in vesicle recycling capability [72]. An increase in synaptophysin levels along with SNAP-25 is also observed in Brodmann's area in the post-mortem brain of patients with bipolar disorder I [73]. Protein levels of synaptotagmin and syntaxin 1 are highly upregulated in areas of cerebral ischemia, which are known to exhibit highly active levels of neurotransmission [74]. Mutations in certain regions of the protein syntaxin 1A, such as Ca^{+2} channel-binding region, are known to increase neurotransmitter release, which suggests that syntaxin 1A is involved in regulating Ca^{+2} channel function [75]. A point mutation in syntaxin 1A is known to result in augmented release of neurotransmitters in models of the fruit fly *Drosophila melanogaster* [76]. Synaptotagmin similar to syntaxin has Ca^{+2} binding domains and is known to form dimers and interact with syntaxin to form complexes in the presence of calcium [77]. Binding of synaptotagmin to SNAP 25 is also calcium dependent [78]. Reticulons are proteins which contribute to lipid membrane curvature and are found in the neuronal porosome complex. The involvement of reticulons with the porosome and diseases associated with their deregulation lend credence to the role of membrane curvature in the regulation of synaptic vesicle fusion at the porosome complex. These studies reflect on the critical role of various neuronal porosome proteins in protein–protein and protein–lipid interactions within the porosome complex and in their participation in normal porosome-mediated neurotransmission at the nerve terminal and in disease.

In addition to the above studies discussed on the involvement of porosome proteins in neurotransmission-related disease, recent morphological studies on the neuronal porosome complex have also shed light on both the health and disease status of the neuronal porosome complex as examined using high-resolution EM [79–83]. Ultrastructure of the neuronal porosome complex in rats subjected to continuous white noise that is relevant to the increasing random noise encountered by humans in today's environment is known to provoke diverse effects on different brain regions [84–89] such as the documented alteration in the length of the porosome complex [83]. Constant exposure to such noise is further known to sabotages the development and normal function of audition, impair hearing, language acquisition [84, 85], memory performance [86, 87], and other cognitive functions.

References

1. Jena, B. P. (2015). 'Porosome' discovered nearly 20 years ago provides molecular insights into the kiss-and-run mechanism of cell secretion. *Journal of Cellular and Molecular Medicine, 19*, 1427–1440. https://doi.org/10.1111/jcmm.12598. PMID 26033351.
2. Naik, A. R., Lewis, K. T., & Jena, B. P. (2015). The neuronal porosome complex in health and disease. *Experimental Biology and Medicine (Maywood, N.J.), 241*(2), 115–130. https://doi.org/10.1177/1535370215598400. PMID 26264442.
3. Hou, X., Lewis, K. T., Wu, Q., Wang, S., Chen, X., Flack, A., Mao, G., Taatjes, D. J., Sun, F., & Jena, B. P. (2013). Proteome of the porosome complex in human airway epithelia: Interaction with the cystic fibrosis transmembrane conductance regulator (CFTR). *Journal of Proteomics, 96*, 82–91. https://doi.org/10.1016/j.jprot.2013.10.041. PMID 24220302.
4. Ceccarelli, B., Hurlbut, W. P., & Mauro, A. (1973). Turnover of transmitter and synaptic vesicles at the frog neuromuscular junction. *The Journal of Cell Biology, 57*, 499–524.
5. Almers, W., & Tse, F. W. (1990). Transmitter release from synapses: Does a preassembled fusion pore initiate exocytosis? *Neuron, 4*, 813–818.
6. Monck, J. R., & Fernandez, J. M. (1992). The exocytotic fusion pore. *The Journal of Cell Biology, 119*, 1395–1404.
7. Neher, E. (1993). Secretion without full fusion. *Nature, 363*, 497–498.
8. Schneider, S. W., Sritharan, K. C., Geibel, J. P., Oberleithner, H., & Jena, B. P. (1997). Surface dynamics in living acinar cells imaged by atomic force microscopy: Identification of plasma membrane structures involved in exocytosis. *Proceedings of the National Academy of Sciences of the United States of America, 94*, 316–321.
9. Cho, S.-J., Quinn, A. S., Stromer, M. H., Dash, S., Cho, J., Taatjes, D. J., & Jena, B. P. (2002). Structure and dynamics of the fusion pore in live cells. *Cell Biology International, 26*, 35–42.
10. Jena, B. P., Cho, S.-J., Jeremic, A., Stromer, M. H., & Abu-Hamdah, R. (2003). Structure and composition of the fusion pore. *Biophysical Journal, 84*, 1–7.
11. Cho, S.-J., Wakade, A., Pappas, G. D., & Jena, B. P. (2002). New structure involved in transient membrane fusion and exocytosis. *Annals of the New York Academy of Sciences, 971*, 254–256.
12. Cho, S.-J., Jeftinija, K., Glavaski, A., Jeftinija, S., Jena, B. P., & Anderson, L. L. (2002). Structure and dynamics of the fusion pores in live GH-secreting cells revealed using atomic force microscopy. *Endocrinology, 143*, 1144–1148.

13. Jeremic, A., Kelly, M., Cho, S.-J., Stromer, M. H., & Jena, B. P. (2003). Reconstituted fusion pore. *Biophysical Journal, 85*, 2035–2043.
14. Naik, A. R., Kulkarni, S. P., Lewis, K. T., Taatjes, D. J., & Jena, B. P. (2016). Functional reconstitution of the porosome complex in live cells. *Endocrinology, 157*, 54–60.
15. Rajagopal, A., Kulkarni, S., Lewis, K. T., Chen, X., Maarouf, A., Kelly, C. V., Taatjes, D. J., & Jena, B. P. (2015). Proteome of the insulin-secreting Min6 porosome complex: Involvement of Hsp90 in its assembly and function. *Journal of Proteomics, 114*, 83–92.
16. Cho, W. J., Jeremic, A., Rognlien, K. T., Zhvania, M. G., Lazrishvili, I., Tamar, B., & Jena, B. P. (2004). Structure, isolation, composition and reconstitution of the neuronal fusion pore. *Cell Biology International, 28*, 699–708.
17. Cho, W. J., Jeremic, A., Jin, H., Ren, G., & Jena, B. P. (2007). Neuronal fusion pore assembly requires membrane cholesterol. *Cell Biology International, 31*, 1301–1308.
18. Cho, W. J., Ren, G., & Jena, B. P. (2008). EM 3D contour maps provide protein assembly at the nanoscale within the neuronal porosome complex. *Journal of Microscopy, 232*, 106–111.
19. Kovari, L. C., Brunzelle, J. S., Lewis, K. T., Cho, W. J., Lee, J.-S., Taatjes, D. J., & Jena, B. P. (2014). X-ray solution structure of the native neuronal porosome-synaptic vesicle complex: Implication in neurotransmitter release. *Micron, 56*, 37–43.
20. Lee, J.-S., Jeremic, A., Shin, L., Cho, W. J., Chen, X., & Jena, B. P. (2012). Neuronal Porosome proteome: Molecular dynamics and architecture. *Journal of Proteomics, 75*(3), 952–962.
21. Lee, J. S., Cho, W. J., Jeftinija, K., Jeftinija, S., & Jena, B. P. (2009). Porosome in astrocytes. *Journal of Cellular and Molecular Medicine, 13*, 365–372. https://doi.org/10.1111/j.1582-4934.2008.00334.x.
22. Cho, W. J., Ren, G., Lee, J. S., Jeftinija, K., Jeftinija, S., & Jena, B. P. (2009). Nanoscale 3D contour map of protein assembly within the astrocyte porosome complex. *Cell Biology International, 33*, 224–229. https://doi.org/10.1016/j.cellbi.2008.11.008. PMID 19084606.
23. Oyler, G. A., Higgins, G. A., Hart, R. A., Battenberg, E., Billingsley, M., Bloom, F. E., & Wilson, M. C. (1989). The identification of a novel synaptosomal-associated protein, SNAP-25, differentially expressed by neuronal subpopulations. *The Journal of Cell Biology, 109*, 3039–3052.
24. Bennett, M. K., Calakos, N., & Schller, R. H. (1992). Syntaxin: A synaptic protein implicated in docking of synaptic vesicles at presynaptic active zones. *Science, 257*, 255–259.
25. Trimble, W. S., Cowan, D. W., & Scheller, R. H. (1988). VAMP-1: A synaptic vesicle-associated integral membrane protein. *Proceedings of the National Academy of Sciences of the United States of America, 85*, 4538–4542.
26. Jeremic, A., Kelly, M., Cho, J.-H., Cho, S.-J., Horber, J. K. H., & Jena, B. P. (2004). Calcium drives fusion of SNARE-apposed bilayers. *Cell Biology International, 28*, 19–31.
27. Jeremic, A., Cho, W. J., & Jena, B. P. (2004). Membrane fusion: What may transpire at the atomic level. *Journal of Biological Physics and Chemistry, 4*, 139–142.
28. Potoff, J. J., Issa, Z., Manke, C. W., Jr., & Jena, B. P. (2008). Ca^{2+}-Dimethylphosphate complex formation: Providing insight into Ca^{2+} mediated local dehydration and membrane fusion in cells. *Cell Biology International, 32*, 361–366.
29. Issa, Z., Manke, C. W., Jr., Jena, B. P., & Potoff, J. J. (2010). Ca^{2+} bridging of apposed phospholipid bilayer. *The Journal of Physical Chemistry. B, 114*, 13249–13254.
30. Ludtke, S. J., Baldwin, P. R., & Chiu, W. (1999). EMAN: Semiautomated software for high-resolution single-particle reconstructions. *Journal of Structural Biology, 128*, 82–97.
31. Frank, J., Radermacher, M., Penczek, P., Zhu, J., Li, Y., Lasjadj, M., & Leith, A. (1996). SPIDER and WEB: Processing and visualization of images in 3D electron microscopy and related fields. *Journal of Structural Biology, 116*, 190–199.
32. Goddard, T. D., Huang, C. C., & Ferrin, T. E. (2005). Software extensions to UCSF chimera for interactive visualization of large molecular assemblies. *Structure, 13*, 473–482.

33. Pettersen, E. F., Goddard, T. D., Huang, C. C., Couch, G. S., Greenblatt, D. M., Meng, E. C., & Ferrin, T. E. (2004). UCSF chimera - a visualization system for exploratory research and analysis. *Journal of Computational Chemistry, 25*, 1605–1612.
34. Cho, W.-J., Shin, L., Ren, G., & Jena, B. P. (2009). Structure of membrane-associated neuronal SNARE complex: Implication in neurotransmitter release. *Journal of Cellular and Molecular Medicine, 13*, 4161–4165.
35. Mohrmann, R., de Wit, H., Verhage, M., Neher, E., & Sørensen, J. B. (2010). Fast vesicle fusion in living cells requires atleast three SNARE complexes. *Science, 330*, 502–505.
36. Cho, W. J., Jeremic, A., & Jena, B. P. (2005). Direct interaction between SNAP-23 and L-type calcium channel. *Journal of Cellular and Molecular Medicine, 9*, 380–386.
37. Cho, W.-J., Lee, J.-S., & Jena, B. P. (2010). Conformation states of the neuronal Porosome complex. *Cell Biology International, 34*, 1129–1132.
38. Lewis, K. T., Maddipati, K. R., Taatjes, D. J., & Jena, B. P. (2014). Neuronal Porosome Lipidome. *Journal of Cellular and Molecular Medicine, 18*, 1927–1937.
39. Nemhauser, I., & Goldberg, D. J. (1985). Structural effects in axoplasm of DNase I, an actin depolymerizer that blocks fast axonal transport. *Brain Research, 334*, 47–58.
40. Cole, J. C., Villa, B. R., & Wilkinson, R. S. (2000). Disruption of actin impedes transmitter release in snake motor terminals. *The Journal of Physiology, 525*(3), 579–586.
41. Klein, M. E., Younts, T. J., Castillo, P. E., & Jordan, B. A. (2013). RNA-binding protein Sam68 controls synapse number and local beta-actin mRNA metabolism in dendrites. *Proceedings of the National Academy of Sciences USA, 110*, 3125–3130.
42. Khanna, R., Zougman, A., & Stanley, E. F. (2007). A proteomic screen for presynaptic terminal N-type calcium channel (CaV2.2) binding partners. *Journal of Biochemistry and Molecular Biology, 40*, 302–314.
43. Khanna, R., Li, Q., Bewersdorf, J., & Stanley, E. F. (2007). The presynaptic CaV2.2 channel-transmitter release site core complex. *The European Journal of Neuroscience, 26*, 547–559.
44. Balestrino, M., Young, J., & Aitken, P. (1999). Block of (Na+,K+)ATPase with ouabain induces spreading depression-like depolarization in hippocampal slices. *Brain Research, 838*, 37–44.
45. Li, K. C., et al. (2011). Follistatin-like 1 suppresses sensory afferent transmission by activating Na+,K+-ATPase. *Neuron, 69*, 974–987.
46. Scuri, R., Lombardo, P., Cataldo, E., Ristori, C., & Brunelli, M. (2007). Inhibition of Na+/K+ ATPase potentiates synaptic transmission in tactile sensory neurons of the leech. *The European Journal of Neuroscience, 25*, 59–167.
47. Kim, J. H., Sizov, I., Dobretsov, M., & von Gersdorff, H. (2007). Presynaptic Ca2+ buffers control the strength of a fast post-tetanic hyperpolarization mediated by the alpha3 Na(+)/K(+)-ATPase. *Nature Neuroscience, 10*, 196–205.
48. Jensen, T. P., Filoteo, A. G., Knopfel, T., & Empson, R. M. (2007). Presynaptic plasma membrane Ca2+ ATPase isoform 2a regulates excitatory synaptic transmission in rat hippocampal CA3. *The Journal of Physiology, 579*, 85–99.
49. Garside, M. L., et al. (2009). Molecular interactions of the plasma membrane calcium ATPase 2 at pre- and post-synaptic sites in rat cerebellum. *Neuroscience, 162*, 383–395.
50. Geerlings, A., Nunez, E., Lopez-Corcuera, B., & Aragon, C. (2001). Calcium- and syntaxin 1-mediated trafficking of the neuronal glycine transporter GLYT2. *The Journal of Biological Chemistry, 276*, 17584–17590.
51. de Juan-Sanz, J., et al. (2013). Na+/K+-ATPase is a new interacting partner for the neuronal glycine transporter GlyT2 that downregulates its expression in vitro and in vivo. *The Journal of Neuroscience, 33*, 14269–14281.
52. Empson, R. M., Akemann, W., & Knopfel, T. (2010). The role of the calcium transporter protein plasma membrane calcium ATPase PMCA2 in cerebellar Purkinje neuron function. *Functional Neurology, 25*, 153–158.
53. Dodson, H. C., & Charalabapoulou, M. (2001). PMCA2 mutation causes structural changes in the auditory system in deafwaddler mice. *Journal of Neurocytology, 30*, 281–292.

54. Iino, S., Kobayashi, S., & Maekawa, S. (1999). Immunohistochemical localization of a novel acidic calmodulin-binding protein, NAP-22, in the rat brain. *Neuroscience, 91*, 1435–1444.
55. Iino, S., & Maekawa, S. (1999). Immunohistochemical demonstration of a neuronal calmodulin-binding protein, NAP-22, in the rat spinal cord. *Brain Research, 834*, 66–73.
56. Yamamoto, Y., Sokawa, Y., & Maekawa, S. (1997). Biochemical evidence for the presence of NAP-22, a novel acidic calmodulin binding protein, in the synaptic vesicles of rat brain. *Neuroscience Letters, 224*, 127–130.
57. Freeman, N. L., & Field, J. (2000). Mammalian homolog of the yeast cyclase associated protein, CAP/Srv2p, regulates actin filament assembly. *Cell Motility and the Cytoskeleton, 45*, 106–120.
58. Zhang, H., Ghai, P., Wang, C., Field, J., & Zhou, G. L. (2013). Mammalian adenylyl cyclase-associated protein 1 (CAP1) regulates cofilin function, the actin cytoskeleton, and cell adhesion. *The Journal of Biological Chemistry, 288*, 20966–20977.
59. Sultana, R., Boyd-Kimball, D., Cai, J., Pierce, W. M., Klein, J. B., Merchant, M., & Butterfield, D. A. (2007). Proteomics analysis of the Alzheimer's disease hippocampal proteome. *Journal of Alzheimer's Disease, 11*, 153–164.
60. Vlkolinsky, R., Cairns, N., Fountoulakis, M., & Lubec, G. (2001). Decreased brain levels of 2′,3′-cyclic nucleotide-3′-phosphodiesterase in down syndrome and Alzheimer's disease. *Neurobiology of Aging, 22*, 547–553.
61. Reinikainen, K. J., Pitkanen, A., & Riekkinen, P. J. (1989). 2′,3′-cyclic nucleotide-3′-phosphodiesterase activity as an index of myelin in the post-mortem brains of patients with Alzheimer's disease. *Neuroscience Letters, 106*, 229–232.
62. Flynn, S. W., et al. (2003). Abnormalities of myelination in schizophrenia detected in vivo with MRI, and post-mortem with analysis of oligodendrocyte proteins. *Molecular Psychiatry, 8*, 811–820.
63. Peirce, T. R., et al. (2006). Convergent evidence for 2′,3′-cyclic nucleotide 3′-phosphodiesterase as a possible susceptibility gene for schizophrenia. *Archives of General Psychiatry, 63*, 18–24.
64. Mikoshiba, K., Aoki, E., & Tsukada, Y. (1980). 2′-3′-cyclic nucleotide 3′-phosphohydrolase activity in the central nervous system of a myelin deficient mutant (Shiverer). *Brain Research, 192*, 195–204.
65. Zhao, Y. Y., Shi, X. Y., Qui, X., Zhang, L., Lu, W., Yang, S., Lee, C., Cheng, G. H., Yang, Z. W., & Tang, Y. (2011). Enriched environment increases the total number of CNPase positive cells in the corpus callosum of middle-aged rats. *Acta Neurobiologiae Experimentalis, 71*, 322–330.
66. Sinclair, L. I., Tayler, H. M., & Love, S. (2015). Synaptic protein levels altered in vascular dementia. *Neuropathology and Applied Neurobiology, 41*, 533–543.
67. Mukaetova-Ladinska, E. B., Xuereb, J. H., Garcia-Sierra, F., Hurt, J., Gertz, H. J., Hills, R., Brayne, C., Huppert, F. A., Paykel, E. S., McGee, M. A., Jakes, R., Honer, W. G., Harrington, C. R., & Wischik, C. M. (2009). Lewy body variant of Alzheimer's disease: Selective neocortical loss of t-SNARE proteins and loss of MAP 2 and alpha-synuclein in medial temporal lobe. *The Scientific World Journal, 9*, 1463–1475.
68. Greber, S., Lubec, G., Cairns, N., & Fountoulakis, M. (1999). Decreased levels of synaptosomal associated protein 25 in the brain of patients with down syndrome and Alzheimer's disease. *Electrophoresis, 20*, 928–934.
69. Corradini, I., Donzelli, A., Antonucci, F., Welzl, H., Loos, M., Martucci, R., De Astis, S., Pattini, L., Inverardi, F., Wolfer, D., Caleo, M., Bozzi, Y., Verderio, C., Frassoni, C., Braida, D., Clerici, M., Lipp, H. P., Sala, M., & Matteoli, M. (2014). Epileptiform activity and cognitive deficits in SNAP-25 (+/−) mice are normalized by antiepileptic drugs. *Cerebral Cortex, 24*, 364–376.
70. McKee, A. G., Loscher, J. S., O'Sullivan, N. C., Chadderton, N., Palfi, A., Batti, L., Sheridan, G. K., O'Shea, S., Moran, M., McCabe, O., Fernandez, A. B., Pangalos, M. N., O'Connor, J. J., Regan, C. M., O'Connor, W. T., Humphries, P., Farrar, G. J., & Murphy, K. J. (2010).

AAV-mediated chronic over-expression of SNAP-25 in adult rat dorsal hippocampus impairs memory-associated synaptic plasticity. *Journal of Neurochemistry, 112*, 991–1004.
71. Smith, R., Klein, P., Koc-Schmitz, Y., Waldvogel, H. J., Faull, R. L., Brundin, P., Plomann, M., & Li, J. Y. (2007). Loss of SNAP-25 and rabphilin 3a in sensory-motor cortex in Huntington's disease. *Journal of Neurochemistry, 103*, 115–123.
72. Jeans, A. F., Oliver, P. L., Johnson, R., Capogna, M., Vikman, J., Molnar, Z., Babbs, A., Partridge, C. J., Salehi, A., Bengtsson, M., Eliasson, L., Rorsman, P., & Davies, K. E. (2007). A dominant mutation in Snap25 causes impaired vesicle trafficking, sensorimotor gating, and ataxia in the blind-drunk mouse. *Proceedings of the National Academy of Sciences of the United States of America, 104*, 2431–2436.
73. Scarr, E., Gray, L., Keriakous, D., Robinson, P. J., & Dean, B. (2006). Increased levels of SNAP-25 and synaptophysin in the dorsolateral prefrontal cortex in bipolar I disorder. *Bipolar Disorders, 8*, 133–143.
74. Cao, F., Hata, R., Zhu, P., Niinobe, M., & Sakanaka, M. (2009). Up-regulation of syntaxin1 in ischemic cortex after permanent focal ischemia in rats. *Brain Research, 1272*, 52–61.
75. Wu, M. N., Fergestad, T., Lloyd, T. E., He, Y., Broadie, K., & Bellen, H. J. (1999). Syntaxin 1A interacts with multiple exocytic proteins to regulate neurotransmitter release in vivo. *Neuron, 23*, 593–605.
76. Lagow, R. D., Bao, H., Cohen, E. N., Daniels, R. W., Zuzek, A., Williams, W. H., Macleod, G. T., Sutton, R. B., & Zhang, B. (2007). Modification of a hydrophobic layer by a point mutation in syntaxin 1A regulates the rate of synaptic vesicle fusion. *PLoS Biology, 5*, e72.
77. Chapman, E. R., An, S., Edwardson, J. M., & Jahn, R. (1996). A novel function for the second C2 domain of synaptotagmin. Ca2+–triggered dimerization. *The Journal of Biological Chemistry, 271*, 5844–5849.
78. Zhang, X., Kim-Miller, M. J., Fukuda, M., Kowalchyk, J. A., & Martin, T. F. (2002). Ca2+–dependent synaptotagmin binding to SNAP-25 is essential for Ca2+–triggered exocytosis. *Neuron, 34*, 599–611.
79. Japaridze, N. J., Okuneva, V. G., Qsovreli, M. G., Surmava, A. G., Lordkipanidze, T. G., Kiladze, M., & Zhvania, M. G. (2012). Hypokinetic stress and neuronal porosome complex in the rat brain: The electron microscopic study. *Micron, 43*(9), 948–963.
80. Okuneva, V. G., Japarodze, N. J., Kotaria, N. T., & Zhvania, M. G. (2012). Neuronal porosome in the rat and cat brain. *Tsitologiia, 54*, 210–215.
81. Zhvania, M.G., Japaridze, N.J., Qsovreli, M.G., Okuneva, V.G., Surmava, A.G., Lordkipanidze, T.G. (2014) The neuronal porosome complex in mammalian brain: A study using electron microscope. NanoCellBiology: Multimodal imaging in biology and medicine. Bhanu P Jena and Douglas J Taajes. Pan Stanford Publishing Pre. Ltd. Boca Raton, FL.
82. Zhvania, M. G., Japaridze, N. J., & Ksovreli, M. (2014). Effect of different forms of hypokinesia on the ultrastructure of limbic, extrapyramidal and neocortical areas of the rat brain: Electron microscopic study. *Proceedings of Physics, 15*, 21–16. Springer International Publishing Switzerland.
83. Zhvania, M. G., Bikashvili, T. Z., Japaridze, N. J., Lazrishvili, I. I., & Ksovreli, M. (2014). White noise and neuronal porosome complex: Transmission electron microscopic study. *Discoveries Jul-Sep, 2*(3), e25.
84. Eggermont, J. J. (2008). The role of sound in adult and developmental auditory cortical plasticity. *Ear and Hearing, 6*, 819–829.
85. Zhu, X., Wang, F., Hu, H., Sun, X., Kilgard, M. P., Merzenich, M. M., & Zhou, X. (2014). Environmental acoustic enrichment promotes recovery from developmentally degraded auditory cortical processing. *The Journal of Neuroscience, 34*(16), 5406–5415.
86. Soderlind, G. B. W., Sikstrom, S., & Sonuga-Narke, E. J. (2010). The effects of background white noise on memory performance in inattentive school children. *Behavioral and Brain Functions, 6*, 55–55.

87. Cheng, L., Wang, S. H., Chen, Q. C., & Liao, X. M. (2011). Moderate noise induced cognition impairment of mice and its underlying mechanisms. *Physiology & Behavior, 104*(5), 981–988.
88. Rausch, V. H., Bauch, E. M., & Bunzeck, N. (2014). White noise improves learning by modulating activity in dopaminergic midbrain regions and right superior temporal sulcus. *Journal of Cognitive Neuroscience, 7*, 1469–1480.
89. Wang, X., Jen, P. H., Wu, F. J., & Chen, Q. C. (2007). Preceding weak noise sharpens the frequency tuning and elevates the response threshold of the mouse inferior collicular neurons through GABAergic inhibition. *Brain Research, 1167*, 80–91.
90. Szklarczyk, D., Franceschini, A., Kuhn, M., Simonovic, M., Roth, A., Minguez, P., et al. (2011). The STRING database in 2011: Functional interaction networks of proteins, globally integrated and scored. *Nucleic Acids Research, 39*, D561–D568.
91. Shin, L., Cho, W.-J., Cook, J., Stemmler, T., & Jena, B. P. (2010). Membrane lipids influence protein complex assembly-disassembly. *Journal of the American Chemical Society, 132*, 5596–5597.
92. de Jong, A. P. H., Roggero, C. M., Ho, M. R., Wong, M. Y., Brautigam, C. A., Rizo, J., & Kaeser, P. S. (2018). RRIM C2B domains target presynaptic active zone functions to PIP2-containing membranes. *Neuron, 98*, 335–349.
93. Sutton, R. B., Fasshauer, D., Jahn, R., & Brunger, A. T. (1998). Crystal structure of a SNARE complex involved in synaptic exocytosis at 2.4 Å resolution. *Nature, 395*, 347–353.
94. Cho, S.-J., Kelly, M., Rognlien, K. T., Cho, J. A., Horber, J. K. H., & Jena, B. P. (2002). SNAREs in opposing bilayers interact in a circular array to form conducting pores. *Biophysical Journal, 83*, 2522–2527.
95. Cho, W.-J., Jeremic, A., & Jena, B. P. (2005). Size of supramolecular SNARE complex: Membrane-directed self-assembly. *Journal of the American Chemical Society, 127*, 10156–10157.
96. Jeremic, A., Quinn, A. S., Cho, W.-J., Taatjes, D. J., & Jena, B. P. (2006). Energy-dependent disassembly of self-assembled SNARE complex: Observation at nanometer resolution using atomic force microscopy. *Journal of the American Chemical Society, 128*, 26–27.
97. Cook, J. D., Cho, W. J., Stemmler, T. L., & Jena, B. P. (2008). Circular dichroism (CD) spectroscopy of the assembly and disassembly of SNAREs: The proteins involved in membrane fusion in cells. *Chemical Physics Letters, 462*, 6–9.
98. Cho, W. J., Lee, J.-S., Ren, G., Zhang, L., Shin, L., Manke, C. W., Potoff, J., Kotaria, N., Zhvania, M. G., & Jena, B. P. (2011). Membrane-directed molecular assembly of the neuronal SNARE complex. *Journal of Cellular and Molecular Medicine, 15*, 31–37.
99. Weber, T., Zemelman, B. V., McNew, J. A., Westerman, B., Gmachi, M., Parlati, F., Söllner, T. H., & Rothman, J. E. (1998). SNAREpins: Minimal machinery for membrane fusion. *Cell, 92*, 759–772.
100. Megighian, A., Scorzeto, M., Zanini, D., Pantano, S., Rossetto, O., Montecucco, C., & Zordan, M. (2010). Arg206 of SNAP-25 is essential for neuroexocytosis at the Drosophila melanogaster neuromuscular junction. *Journal of Cell Science, 123*, 3276–3283.
101. Hammel, I., & Meilijson, I. (2013). Function suggests nano-structure: Toward a unifying theory for secretion rate, a statistical mechanics approach. *Journal of Royal Society Interface, 10*, 20130640. https://doi.org/10.1098/rsif.2013.0640.
102. Alverez de Toledo, G., Fernandez-Chacon, R., & Fernandez, J. M. (1991). Release of vesicle products during transient vesicle fusion. *Nature, 363*, 554–558.
103. Monck, J. R., Oberhauser, A. F., Alvarez de Toledo, G., & Fernandez, J. M. (1991). Is swelling of the secretory granule matrix the force that dilates the exocytotic fusion pore? *Biophysical Journal, 59*, 39–47.
104. Jena, B. P., Schneider, S. W., Geibel, J. P., Webster, P., Oberleithner, H., & Sritharan, K. C. (1997). Gi regulation of secretory vesicle swelling examined by atomic force microscopy. *Proceedings of the National Academy of Sciences of the United States of America, 94*, 13317–13322.

105. Kelly, M., Cho, W.-J., Jeremic, A., Abu-Hamdah, R., & Jena, B. P. (2004). Vesicle swelling regulates content expulsion during secretion. *Cell Biology International, 28*, 709–716.
106. Cho, S.-J., Sattar, A. K. M., Jeong, E.-H., Satchi, M., Cho, J., Dash, S., Mayes, M. S., Stromer, M. H., & Jena, B. P. (2002). Aquaporin 1 regulates GTP-induced rapid gating of water in secretory vesicles. *Proceedings of the National Academy of Sciences of the United States of America, 99*, 4720–4724.
107. Jeremic, A., Cho, W.-J., & Jena, B. P. (2005). Involvement of water channels in synaptic vesicle swelling. *Experimental Biology and Medicine, 230*, 674–680.
108. Shin, L., Basi, N., Lee, J.-S., Cho, W.-J., Chen, Z., Abu-Hamdah, R., Oupicky, D., & Jena, B. P. (2010). Involvement of vH^+-ATPase in synaptic vesicle swelling. *Journal of Neuroscience Research, 88*, 95–101.
109. Lee, J.-S., Cho, W.-J., Shin, L., & Jena, B. P. (2010). Involvement of cholesterol in synaptic vesicle swelling. *Experimental Biology and Medicine, 235*, 470–477.
110. Chen, Z.-H., Lee, J.-S., Shin, L., Cho, W.-J., & Jena, B. P. (2011). Involvement of β-adrenergic receptor in synaptic vesicle swelling and implication in neurotransmitter release. *Journal of Cellular and Molecular Medicine, 15*, 572–576.
111. Kelly, M., Abu-Hamdah, R., Cho, S.-J., Ilie, A. L., & Jena, B. P. (2005). Patch clamped single pancreatic zymogen granules: Direct measurement of ion channel activities at the granule membrane. *Pancreatology, 5*, 443–449.

Chapter 2
Ubiquitin–Proteasome Machinery: Cells Garbage Disposal

Abstract Proteins are cells building blocks with a highly dynamic existence. Rates of synthesis and degradation dictate cellular protein levels at any given moment. Many proteins such as transcription factors or signaling molecules are rapidly degraded (min), while structural proteins have very slow turnover rates (days). Similarly, damaged or improperly folded proteins are rapidly degraded. Two major cellular pathways operate to mediate the degradation of cellular proteins. One involves the ATP-dependent ubiquitin–proteasome pathway and the other involves the non-ATP-dependent proteolysis that occurs in the lysosome. In this chapter, the discovery, structure–function, and disease resulting from impaired ubiquitin–-proteasome-mediated protein degradation are discussed.

Keywords Proteasome · Ubiquitin · ATP-dependent degradation · Cells garbage disposal

B. P. Jena, *Cellular Nanomachines*,
https://doi.org/10.1007/978-3-030-44496-9_2

Discovery

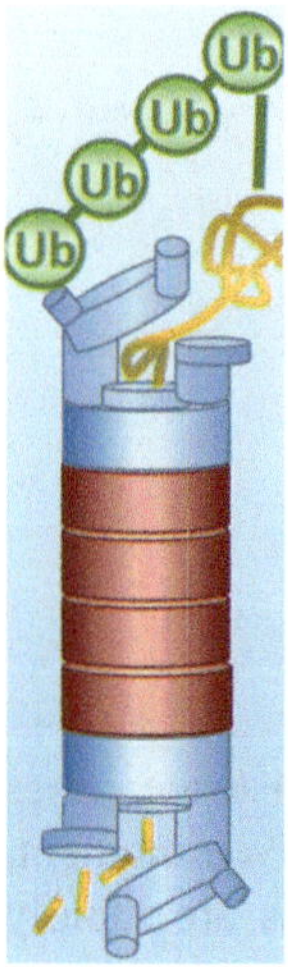

The use of adenosine triphosphate (ATP) in majority (>80%) of cellular protein degradation processes had been proposed since the 1950s. However, only after the publication of a 1971 study [1] by *Avram Hershko and Gordon M. Tomkins*, did the proposed hypothesis firmly established the requirement of ATP on cellular protein degradation. In this study on degradation of tyrosine aminotransferase in cultured hepatoma cells, it was reported that sodium fluoride and other inhibitors of cellular ATP production inhibited the degradation of tyrosine aminotransferase and other proteins [1]. This study [1] conclusively demonstrated for the first time, the requirement of ATP in protein degradation in cells, unraveling a new cellular paradigm. Then in 1978, in a seminal in vitro study [2] by *Aaron Ciechanover, Y. Hod*, and *Avram Hershko*, it was reported that the degradation of denatured globin by two reticulocyte lysate fractions I and II resolved using gel filtration chromatography is ATP-dependent. This work was a breakthrough in our understanding of ATP-mediated protein degradation in cells. Fractions I and II were named in the order of their elution profile from the DEAE-cellulose column. Fraction II had a neutral protease activity, whereas the heat-stable fraction I (later identified as ***ubiquitin***) had no proteolytic activity but restored ATP-dependent proteolysis when combined with fraction II. This 1978 study [2], followed by two seminal works, one reported in 1979 [3] and the other in 1980 [4], firmly established the presence of an ATP-dependent protein degradation machinery in cells. In the 1979 [3] study by *Avram Hershko, Aaron Ciechanover*, and *Irwin A. Rose*, fraction II was further resolved into two fractions, a heat-labile and a heat-stable fraction with complementing activities. The highly heat-labile fraction could only be stabilized by ATP, but not GTP or CTP [3]. It was further demonstrated that although ADP or the nonhydrolyzable β, γ, imido or methyl analogous of ATP stabilized the heat-labile fraction, they could not activate proteolysis. All three components (fraction I and the two sub-fractions of fraction II) and ATP are required for proteolytic activity [3]. Next, in the 1980 study by *Hershko* et al., the conjugation of protein multiple chains of the polypeptide of ATP-dependent proteolysis was demonstrated [4]. It was demonstrated that the heat-stable polypeptide ATP-dependent proteolysis factor I (APF-I) forms covalent compounds with proteins in an ATP-dependent reaction [4]. Removal of ATP following the formation of AFP-I conjugates with proteins, resulted in the regeneration of AFP-I, suggesting the presence of an amidase [4]. This explained the role of ATP in the proteolytic activity. Later studies in 1982 [5] and 1983 [6] identified the enzymes E1 (ubiquitin-activating), E2 (ubiquitin-conjugating), and E3 (ubiquitin-ligating) that polymerize ubiquitin to serve as the degradation signal to shuttle the target proteins to a cylindrical multi-catalytic protease complex (Fig. 2.1) called the *proteasome* [7–12], for degradation. The proteasome structure was initially captured by electron microscopy (EM) [13], followed by studies involving multiple imaging modalities including EM [14–16], X-ray [17], nuclear magnetic resonance (NMR) imaging [18], and atomic force microscopy (AFM) [19].

Proteasome Structure–Function

In all eukaryotic cells, the 26S proteasome catalyzes most of the ATP-dependent intracellular protein degradation (Fig. 2.1). The 26S proteasome is a 2.5 MDa multi-catalytic degradation machine. The protease core of this large protein machine is a barrel-shaped 20S core particle (CP) that is capped at both ends by a 19S regulatory

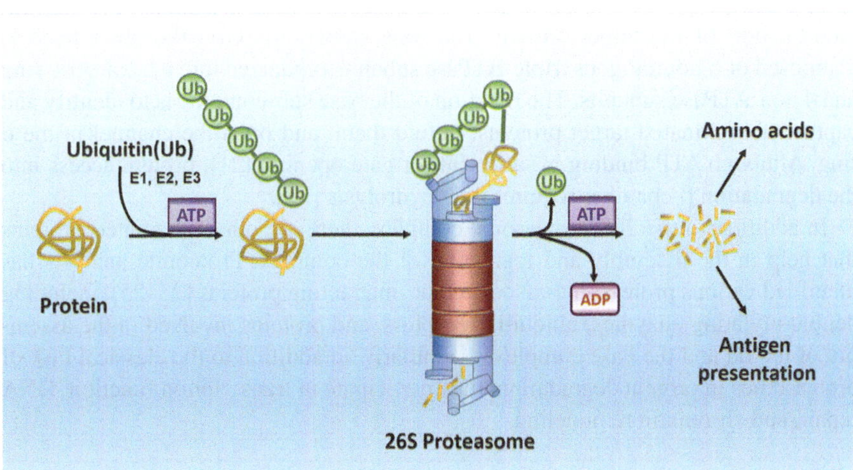

Fig. 2.1 Schematic diagram of the 26S proteasome, the major protein degrading nanomachine in eukaryotic cells. Note the 20S α,β,β,α protease rings forming the central catalytic core particle (CP) colored maroon that is capped at each end by the 19S regulatory particle (RP) in sky blue

particle (RP). The 19S RP also serves to recognize ubiquitinated client proteins. Ubiquitinated proteins targeted for degradation first attach to the 19S RP where they are unfolded, then translocated into the 20S CP for degradation and subsequently released out the other end of the proteasome cylinder. This ubiquitin–proteasome-mediated protein degradation machinery plays a critical role in governing a wide range of cellular functions, from normal protein turnover to the degradation of misfolded and damaged proteins, to control of the cell cycle [20]. The highly conserved cylindrical 20S CP, with a molecular mass of approximately 750 kDa, is formed of axial stacking of two outer α rings and two inner β rings. Each α and β rings are composed of seven structurally similar α and β subunits arranged in an $\alpha_{1-7}\beta_{1-7}\beta_{1-7}\alpha_{1-7}$ configuration. The β-type subunits of each inner ring possess catalytically active threonine residues at their N termini, demonstrating N-terminal hydrolase activity. This indicates that the proteasome is a threonine protease. Similarly, other subunits possess caspase-like peptidyl-glutamyl-peptide hydrolyzing, trypsin- and chymotrypsin-like activities facing the interior of the cylinder, and therefore capable of cleaving peptide bonds at the C-terminal side of acidic, basic, and hydrophobic amino acid residues of subject proteins targeted for degradation. Using this machinery, the 20S CP is able to degrade in a progressive manner, subject proteins first into 3–15 amino acid in length oligopeptides, followed by amino acids using oligo peptidases and amino carboxyl peptidases.

The 19S RP recognizes subject proteins targeted for degradation by their polyubiquitin chain. The RP then removes the polyubiquitin chain from the targeted protein, unfolds the protein, opens the outer α ring, and transfers the protein into the CP for degradation. The RP is composed of 20 subunits classified into two groups namely the regulatory triple-ATPase group and the non-ATPase group. Furthermore, the 19S RP complex has two sub-complexes, the lid and the base. The lid sub-complex is composed of 9 non-ATPase subunits, whose major function is the deubiquitination of the target protein. The base sub-complex on the other hand is composed of 6 homologous triple-ATPase subunits organized into a hexameric ring and 4 non-ATPase subunits. The function of the base sub-complex is to identify and capture ubiquitinated target proteins, unfold them, and open the channel in the α ring. Although ATP binding is sufficient for gate opening [21], protein access into the degradation β-chamber requires ATP hydrolysis [22].

In addition to the 26S proteasome complex, there are other associated proteins that help in the assembly and regulation of the complex. Proteomic analysis has identified various proteasome-associated or -interacting proteins [23–25], including deubiquitylating enzymes, ubiquitin receptors, and proteins involved in the assembly of the lid and the base complexes. Similarly, in addition to the classical task of proteasomes in protein degradation, they participate in transcription function, DNA repair, and chromatin remodeling.

Disease and Treatments Targeting the Proteasome

The highly dynamic cellular protein pool is precisely regulated via the cross-talk between protein synthesis, folding, and degradation machineries. Disturbances in this cross-talk therefore result in disease states. Consequently, the ubiquitin–proteasome system has been implicated in regulation of the cell cycle, in immune and inflammatory responses, and in DNA repair, cancer, and apoptosis, among others. Not surprisingly, alterations in the genes coding for proteasome proteins have been implicated in a wide range of diseases, among them, coronary artery disease [26], type 2 diabetes mellitus [27], and Grave's disease [28]. In the past 15 years, there has been much interest in treating a wide range of diseases by modulating proteasomes function. Bortezomib, a N-protected dipeptide that binds with high affinity and specificity to the catalytic domain of the 26S proteasome to prevent degradation, has been approved by the Food and Drug Administration (FDA) since 2003, and sold under the brand name Velcade, to treat multiple myeloma and mantle cell lymphoma [29, 30]. The potency and efficacy of proteasome inhibition for cancer treatment rely on the role of drugs to regulate cell proliferation and on the reliance of cancer cells on proteasome function. In contrast, diseases such as neurodegenerative diseases are the consequence of the accumulation of toxic misfolded proteins, resulting in the collapse of cross-talk between protein synthesis, folding, and degradation machineries. Therefore, both proteasome inhibition and activation represent potential targets for future drug development.

References

1. Hershko, A., & Tomkins, G. M. (1971). Studies on the degradation of tyrosine aminotransferase in hepatoma cells in culture. *The Journal of Biological Chemistry, 246*, 710–714.
2. Ciechanover, A., Hod, Y., & Hershko, A. (1978). A heat-stable polypeptide component of an ATP-dependent proteolytic system from reticulocytes. *Biochemical and Biophysical Research Communications, 81*, 1100–1105.
3. Hershko, A., Ciechanover, A., & Rose, I. A. (1979). Resolution of the ATP-dependent proteolytic system from reticulocytes: A component that interacts with ATP. *Proceedings of the National Academy of Sciences, 76*, 3107–3110.
4. Hershko, A., Ciechanover, A., Heller, H., Haas, A. L., & Rose, I. A. (1980). Proposed role of ATP in protein breakdown: Conjugation of protein with multiple chains of the polypeptide of ATP-dependent proteolysis. *Proceedings of the National Academy of Sciences, 77*, 1783–1786.
5. Ciechanover, A., Elias, S., Heller, H., & Hershko, A. (1982). "Covalent affinity" purification of ubiquitin-activating enzyme. *The Journal of Biological Chemistry, 257*, 2537–2542.
6. Hershko, A., Heller, H., Elias, S., & Ciechanover, A. (1983). Components of ubiquitin–protein ligase system. Resolution, affinity purification, and role in protein breakdown. *The Journal of Biological Chemistry, 258*, 8206–8214.
7. Wilk, S., & Orlowski, M. (1980). Cation-sensitive neutral endopeptidase: Isolation and specificity of the bovine pituitary enzyme. *Journal of Neurochemistry, 35*, 1172–1182.
8. Wilk, S., & Orlowski, M. (1983). Evidence that pituitary cation-sensitive neutral endopeptidase is a multicatalytic protease complex. *Journal of Neurochemistry, 40*, 842–849.

9. Orlowski, M., & Wilk, S. (1988). Multicatalytic proteinase complex' or 'multicatalytic proteinase': A high-Mr-endopeptidase. *The Biochemical Journal, 255*, 751–751.
10. Rieder, R., Ibrahim, A., & Etlinger, J. (1985). A particle-associated ATP-dependent proteolytic activity in erythroleukemia cells. *The Journal of Biological Chemistry, 260*, 2015–2018.
11. Waxman, L., Fagan, J., & Goldberg, A. L. (1987). Demonstration of two distinct high molecular weight proteases in rabbit reticulocytes, one of which degrades ubiquitin conjugates. *The Journal of Biological Chemistry, 262*, 2451–2457.
12. Azaryan, A., Banay-Schwartz, M., & Lajtha, A. (1989). ATP+ubiquitin-dependent proteinase and multicatalytic proteinase complex in bovine brain. *Neurochemical Research, 14*, 995–1001.
13. Ikai, A., Nishigai, M., Tanaka, K., & Ichihara, A. (1991). Electron microscopy of 26 S complex containing 20 S proteasome. *FEBS Letters, 292*, 21–24.
14. Lowe, J., Stock, D., Jap, B., Zwickl, P., Baumeister, W., & Huber, R. (1995). Crystal structure of the 20S proteasome from the archaeon T. acidophilum at 3.4 a resolution. *Science, 268*, 533–539.
15. Walz, J., Erdmann, A., Kania, M., Typke, D., Koster, A. J., & Baumeister, W. (1998). 26S proteasome structure revealed by three-dimensional electron microscopy. *Journal of Structural Biology, 121*, 19–29.
16. Nickell, S., Beck, F., Korinek, A., Mihalache, O., Baumeister, W., & Plitzko, J. M. (2007). Automated cryoelectron microscopy of "single particles" applied to the 26S proteasome. *FEBS Letters, 581*, 2751–2756.
17. Groll, M., Ditzel, L., Lowe, J., Stock, D., Bochtler, M., Bartunik, H. D., & Huber, R. (1997). Structure of 20S proteasome from yeast at 2.4 a resolution. *Nature, 386*, 463–471.
18. Sprangers, R., & Kay, L. E. (2007). Quantitative dynamics and binding studies of the 20S proteasome by NMR. *Nature, 445*, 618–622.
19. Rosenzweig, R., Osmulski, P. A., Gaczynska, M., & Glickman, M. H. (2008). The central unit within the 19S regulatory particle of the proteasome. *Nature Structural & Molecular Biology, 15*, 573–580.
20. Hershko, A., & Ciechanover, A. (1998). The ubiquitin system. *Annual Review of Biochemistry, 67*, 425–479.
21. Saeki, Y., & Tanaka, K. (2007). Unlocking the proteasome door. *Molecular Cell, 27*, 865–867.
22. Liu, C. W., Li, X., Thompson, D., Wooding, K., Chang, T. L., Tang, Z., Yu, H., Thomas, P. J., & DeMartino, G. N. (2006). ATP binding and ATP hydrolysis play distinct roles in the function of 26S proteasome. *Molecular Cell, 24*, 39–50.
23. Verma, R., Chen, S., Feldman, R., Schieltz, D., Yates, J., Dohmen, J., & Deshaies, R. J. (2000). Proteasomal proteomics: Identification of nucleotide-sensitive proteasome-interacting proteins by mass spectrometric analysis of affinity- purified proteasomes. *Molecular Biology of the Cell, 11*, 3425–3439.
24. Guerrero, C., Tagwerker, C., Kaiser, P., & Huang, L. (2006). An integrated mass spectrometry-based proteomic approach: Quantitative analysis of tandem affinity-purified in vivo cross-linked protein complexes (QTAX) to decipher the 26 S proteasome-interacting network. *Molecular & Cellular Proteomics, 5*, 366–378.
25. Wang, X., Chen, C. F., Baker, P. R., Chen, P. L., Kaiser, P., & Huang, L. (2007). Mass spectrometric characterization of the affinity-purified human 26S proteasome complex. *Biochemistry, 46*, 3553–3565.
26. Alsmadi, O., Muiya, P., Khalak, H., Al-Saud, H., Meyer, B. F., Al-Mohanna, F., Alshahid, M., & Dzimiri, N. (2009). Haplotypes encompassing the KIAA0391 and PSMA6 gene cluster confer a genetic link for myocardial infarction and coronary artery disease. *Annals of Human Genetics, 73*(5), 475–483.
27. Sjakste, T., Kalis, M., Poudziunas, I., Pirags, V., Lazdins, M., Groop, L., & Sjakste, N. (2007). Association of microsatellite polymorphisms of the human 14q13.2 region with type 2 diabetes mellitus in Latvian and Finnish populations. *Annals of Human Genetics, 71*(6), 772–776.

28. Sjakste, T., Eglite, J., Sochnevs, A., et al. (2004). Microsatellite genotyping of chromosome 14q13.2-14q13 in the vicinity of proteasomal gene PSMA6 and association with Graves' disease in the Latvian population. *Immunogenetics, 56*(4), 238–243.
29. Esseltine, D. L., & Mulligan, G. (2012). An historic perspective of proteasome inhibition. *Seminars in Hematology, 49*, 196–206.
30. Hoeller, D., & Dikic, I. (2009). Targeting the ubiquitin system in cancer therapy. *Nature, 458*, 438–444.

Chapter 3
Chaperonin: Protein Folding Machinery in Cells

Abstract As ribosomes initiate synthesis of a polypeptide chain, its precise folding into the functional three-dimensional protein structure is achieved either by spontaneous self-folding or by chaperonin-assisted folding requiring ATP. Spontaneous self-folding is achieved and dictated by its linear amino acid sequence and the existing intracellular environment (such as pH and temperature of the nano environment) to be thermodynamically favorable to assume a negative Gibbs free energy value. In contrast, chaperonin assists folding by preventing incorrect folding conformations and aggregation. Chaperonins are double ring structures stacked one over the other to form a protein-folding chamber in the center of the stack. In *E. coli,* the protein-folding chaperone machinery measures 18.4 nm in length and 14 nm wide and is comprised of two rings stacked one over the other, with each ring composed of seven subunits. Each subunit is composed of three domains, including an ATP-binding domain. The binding of ATP leads to the establishment of a hydrophilic cage for protein folding. Mutations in genes encoding chaperones, or changes in the expression levels of chaperones, have been identified to result in a wide range of disorders.

Keywords Chaperonin · ATP-dependent protein folding · GroEL · GroES

B. P. Jena, *Cellular Nanomachines*,
https://doi.org/10.1007/978-3-030-44496-9_3

Discovery

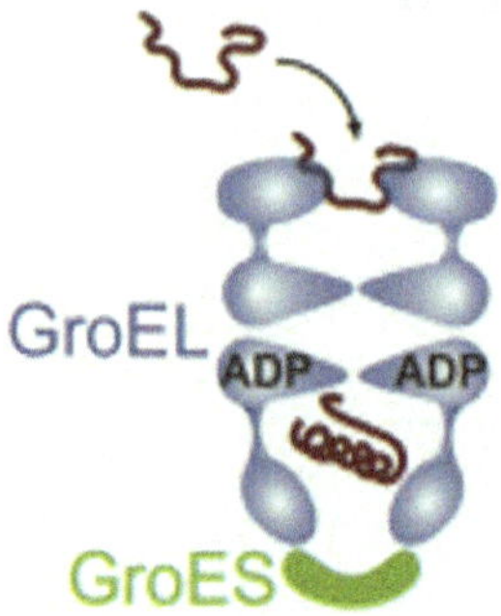

The discovery of the chaperonin protein folding nanomachine in the early 1970s by *Costa P. Georgopoulos* and colleagues [1, 2] and elucidation of its structure at the nanoscale using electron microscopy (EM) by *Roger W. Hendrix* [3] was pivotal in the discovery of chaperonin, the protein folding nanomachine in cells [4]. In the 1979 study [3], *Roger W. Hendrix* reports "….groE is a soluble protein which is found as an oligomer containing 14 subunits of molecular weight 65,000 each. The gp groE particle is cylindrical with a diameter of 125 Å and a height of 100 Å, and it has sevenfold rotational symmetry. It has a weak ATPase activity, …." This was a

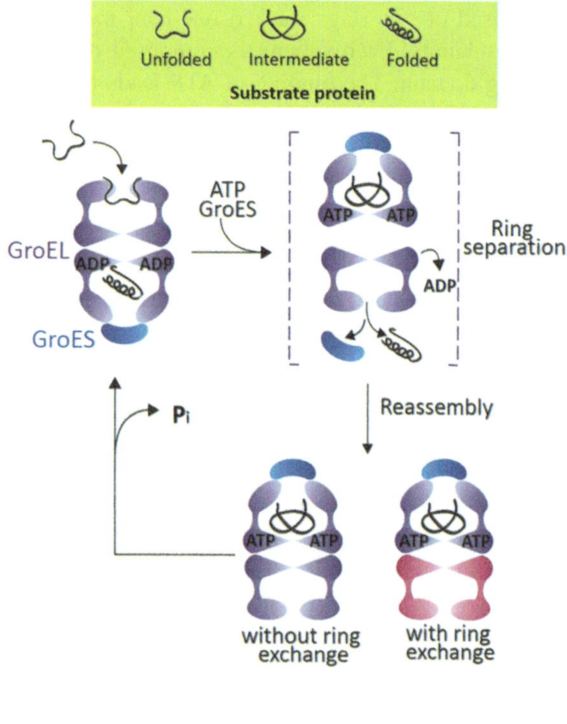

Fig. 3.1 Schematic diagram of chaperonin nanomachine in cells, depicting its ATP-mediated protein folding cycle. The binding of ATP to *GroEL* promotes the binding of the lid-shaped *GroES*, leading to the establishment of a hydrophilic cage for protein folding

discovery made while trying to define the role of the bacterial host (*Escherichia coli*) in the development life cycle of the λ phage virus. Simple yet elegant genetic screens of *Escherichia coli gro* mutants demonstrated that the *E. coli groEL* and *groES* genes encode the GroEL and GroES proteins and are required for the proper assembly of the T4 bacteriophage [1, 2]. The 1979 study [3] utilizing EM and biochemical approaches was the first to report that the *E. coli GroEL* forms complexes composed of two stacked seven-meric rings to establish the cage structure of the chaperone nanomachine. In view of these earlier findings that *GroE* chaperone machines are required for phage morphogenesis and later findings that the eukaryotic *GroEL* homolog Hsp60 is involved in the correct assembly and folding of the large oligomeric photosynthetic CO_2 fixing enzyme RuBisCO (ribulose 1,5-bisphosphate carboxylase/oxygenase) in plants [5–7] laid the foundation that resulted in an explosion of both structural and functional studies [8–10] on the chaperonin protein folding machinery in cells. The demonstration that *GroEL* and *GroES* proteins in the presence of ATP could reconstitute active dimeric RuBisCO was a turning point that suggested these proteins have a very general role in the folding and assembly of cellular proteins [7]. These earlier studies also opened the door to protein import studies in the yeast mitochondria in the late 1980s and early 1990s, providing a wealth of information to further our understanding of the structure and function of the chaperone protein folding machinery in cells (Fig. 3.1) [11–15]. Earlier understanding that polypeptides contain all the information in their amino acid sequence required to fold and reach native form [16] was simply not completely correct. Some proteins do require help to fold and mature in an energy-dependent manner, via the chaperone protein folding machinery (Fig. 3.1), a paradigm shift in our understanding of protein folding in cells.

Structure–Function

The nanoscale structure of chaperonin was first resolved in the late 1970s using EM [3], and nearly two decades later [14, 15] the near atomic structure of the chaperonin nanomachine in its ADP-bound state was solved using X-ray crystallography [15]. Chaperonins are double ring structures stacked one over the other to form a protein-folding chamber in the center of the stack (Fig. 3.1). Chaperonins are subdivided into two groups: Group I and Group II. Group I chaperonins are composed of Hsp60 and its co-chaperone Hsp10. Group I chaperonins are found in bacteria, mitochondria, and chloroplasts, where Hsp10 functions as "lid" for the Hsp60 chaperonin cage. In contrast, Group II chaperonins are found in archaea and in the cytosol of eukaryotes, where instead of a co-chaperone serving as the "lid," subunits of the chaperone themselves serve as the "lid" domain. The most intensely investigated chaperone is the *E. coli GroEL–GroES* protein-folding chaperone machinery. The *E. coli GroEL–GroES* protein-folding chaperone machinery measures 18.4 nm in length and 14 nm wide. The *GroEL* of chaperonin is comprised of two rings stacked one over the other, with each ring composed of seven 57 kDa

GroEL subunits [3, 14]. Each subunit is composed of three domains: an ATP-binding domain, an intermediate hinge domain, and a *GroES*-binding domain. The binding of ATP to *GroEL* promotes the binding of the lid-shaped *GroES*, leading to the establishment of a hydrophilic cage for protein folding (Fig. 3.1). The precise mechanism involved in *GroEL–GroES*-mediated protein folding however is still unclear. Among the proposed mechanisms involved in chaperonin-mediated protein folding in addition to preventing aggregation is that the space constraints within the chaperonin cage. This constraint imposed on the unfolded protein and the interaction of the unfolded proteins with structured water molecules associated with negatively charged residues lining the cage are thought to promote protein folding [17]. It is important to note that since *GroEL–GroES* protein-folding chaperone machinery has a physically limited cage volume, only 30–60 kDa in size proteins could be accommodated for folding [18]. Some larger unfolded proteins, however, have been known to utilize *GroEL* primarily to prevent their aggregation rather than for folding [19, 20].

In addition to protein folding and preventing their aggregation in cells, chaperonins have been demonstrated to participate in the proper assembly of higher order cellular structures. In the assembly of a multi-subunit complex, for example, the individual subunits achieve their native conformation in the context of the final complex. The α-tubulin–β-tubulin heterodimer is a case in point. A number of cytosolic chaperones in precise order associate with the newly translated peptide and direct its conformation to enable binding and hydrolysis of GTP by tubulins to promote heterodimerization [21]. The chaperon prefoldin facilitates the delivery of nascent α and β tubulins to cytosolic chaperons [22]. It is demonstrated that the association of unfolded α tubulin with cytosolic chaperons is critical for folding of the GTP-binding pocket [23]. Once such a semi-native conformation is achieved inside the chaperon "cage," a monomeric nucleotide-bound tubulin is released to interact with other chaperones [24]. This leads to nucleotide exchange and hydrolysis by β tubulin, prior to the establishment of α–β heterodimers [25], where the GTP-binding pocket of α tubulin is buried at the α–β interface [26]. Tubulin α–β heterodimers are then able to polymerize head-to-tail and associate laterally to form microtubules [27]. Similarly, in the assembly of the porosome complex, the critical presence of the late chaperon HSP90 is required [28]. In this study [28] half-maximal inhibition of Hsp90 using the specific inhibitor 17-demethoxy-17-(2-prophenylamino) geldanamycin results in the loss of porosome proteins, including the calcium-transporting ATPase type 2C and the potassium channel subfamily K member 2 from the Min6 porosome complex. This loss of porosome proteins is reflected in the observed inhibition of glucose stimulated insulin release from Min6 cells exposed to the Hsp90 specific inhibitor, implicating Hsp90 in the assembly and function of the insulin-secreting porosome complex.

Disease Resulting from Alterations in Chaperonin

Mutations in genes encoding chaperones, or changes in the expression levels of chaperones, have been identified to result in a wide range of disorders [29–37]. Hereditary spastic paraplegia (HSP), for example, represents a broad range of human neurodegenerative disorders, some of which are caused by defects in paraplegin and spastin, both thought to function as Group I chaperones. Similarly, in McKusick–Kaufman syndrome where patients display heart and other ailments, the altered gene encoding the protein is analogous to a subunit of the Group II archaeal chaperonin [38, 39]. In X-linked retinitis pigmentosa where progressive degeneration of photoreceptive cells in the retina occurs, eventually leading to blindness, one among the five chaperones involved in the folding of tubulin is reported to be impaired via missense and/or protein truncation mutation [40, 41]. Deletion of the gene encoding chaperone protein Hsp27 has been found in patients with Williams syndrome where mental retardation, cytoskeletal abnormalities, facial dysmorphism, and arteriopathy among other disorders are observed [42]. In some cases of Charcot–Marie–Tooth disease characterized by peripheral nerve degeneration, Hsp27 has been found with a missense mutation [43]. Some form of cataracts results from misfolded proteins including crystallins in the lens of the eye [44–48]. These are just a few examples of pathologies linked to altered chaperonin function and resultant unfolded or misfolded proteins.

References

1. Georgopoulos, C. P., Hendrix, R. W., Kaiser, A. D., & Wood, W. B. (1972). Role of the host cell in bacteriophage morphogenesis: Effects of a bacterial mutation on T4 head assembly. *Nature: New Biology, 239*, 38–41.
2. Georgopoulos, C. P., Hendrix, R. W., Casjens, S. R., & Kaiser, A. D. (1973). Host participation in bacteriophage lambda head assembly. *Journal of Molecular Biology, 76*, 45–60.
3. Hendrix, R. W. (1979). Purification and properties of groE, a host protein involved in bacteriophage assembly. *Journal of Molecular Biology, 129*, 375–392.
4. Georgopoulos, C. P. (2006). Toothpicks, serendipity and the Emergence of the Escherichia coli DnaK (Hsp70) and GroEL (Hsp60) Chaperone Machines. Anecdotal, Historical and Critical Commentaries on Genetics Edited by James F. Crow and William F. Dove. *Genetics, 174*, 1699–1707.
5. Hemmingsen, S. M., Woolford, C., van der Vies, S. M., Tilly, K., Dennis, D. T., Georgopoulos, C. P., Hendrix, R. W., & Ellis, R. J. (1988). Homologous plant and bacterial proteins chaperone oligomeric protein assembly. *Nature, 333*, 330–334.
6. Bochkareva, E. S., Lissin, N. M., & Girshovich, A. S. (1988). Transient association of newly synthesized unfolded proteins with the heat-shock GroEL protein. *Nature, 336*, 254–257.
7. Goloubinoff, P., Christeller, J. T., Gatenby, A. A., & Lorimer, G. H. (1989). Reconstitution of active dimeric ribulose bisphosphate carboxylase from an unfolded state depends on two chaperonin proteins and mg-ATP. *Nature, 342*, 884–889.
8. Lindquist, S., & Craig, E. A. (1988). The heat-shock proteins. *Annual Review of Genetics, 22*, 631–677.

9. Bukau, B., & Horwich, A. L. (1998). The Hsp70 and Hsp60 chaperone machines. *Cell, 92*, 351–366.
10. Hartl, F. U., & Hayer-Hartl, M. (2002). Molecular chaperones in the cytosol: From nascent chain to folded protein. *Science, 295*, 1852–1858.
11. Cheng, M. Y., Pollock, R. A., Hendrick, J. P., & Horwich, A. L. (1987). Import and processing of human ornithine transcarbamoylase precursor by mitochondria from *Saccharomyces cerevisiae. Proceedings of the National Academy of Sciences of the United States of America, 84*, 4063–4067.
12. Cheng, M. Y., Hartl, F. U., Martin, J., Pollock, R. A., Kalousek, F., Neupert, W., Hallberg, E. M., Hallberg, R. L., & Horwich, A. L. (1989). Mitochondrial heat-shock protein hsp60 is essential for assembly of proteins imported into yeast mitochondria. *Nature, 337*, 620–625.
13. Weissman, J. S., Rye, H. S., Fenton, W. A., Beechem, J. M., & Horwich, A. L. (1996). Characterization of the active intermediate of a GroEL-GroES-mediated protein folding reaction. *Cell, 84*, 481–490.
14. Braig, K., Otwinowski, Z., Hegde, R., Boisvert, D. C., Joachimiak, A., Horwich, A. L., & Sigler, P. B. (1994). The crystal structure of the bacterial chaperonin GroEL at 2.8 A. *Nature, 371*, 578–586.
15. Xu, Z., Horwich, A. L., & Sigler, P. B. (1997). The crystal structure of the asymmetric GroEL-GroES-(ADP) chaperonin complex. *Nature, 388*, 741–750.
16. Anfinsen, C. B. (1973). Principles that govern the folding of protein chains. *Science, 181*, 223–230.
17. Chakraborty, K., Chatila, M., Sinha, J., Shi, Q., Poschner, B. C., Sikor, M., Jiang, G., Lamb, D. C., Hartl, F. U., & Hayer-Hartl, M. (2010). Chaperonin-catalyzed rescue of kinetically trapped states in protein folding. *Cell, 142*, 112–122.
18. Motojima, F., & Yoshida, M. (2010). Polypeptide in the chaperonin cage. *Cell, 84*, 481–490.
19. Kerner, M. J., Naylor, D. J., Ishihama, Y., Maier, T., Chang, H. C., Stines, A. P., Georgopoulos, C. P., Frishman, D., Hayer-Hartl, M., & Hartl, F. U. (2005). Proteome-wide analysis of chaperonin- dependent protein folding in *Escherichia coli. Cell, 122*, 209–220.
20. Chaudhuri, T. K., Verma, V. K., & Maheswari, A. (2009). GroEL assisted folding of large polypeptide substrates in Escherichia coli: Present scenario and assignments for the future. *Progress in Biophysics and Molecular Biology, 99*, 42–50.
21. Lundin, V. F., Leroux, M. R., & Stirling, P. C. (2010). Quality control of cytoskeletal proteins and human disease. *Trends in Biochemical Sciences, 35*, 288–297.
22. Vainberg, I. E., Lewis, S. A., Rommelaere, H., Ampe, C., Vandekerckhove, J., Klein, H. L., & Cowan, N. J. (1998). Prefoldin, a chaperone that delivers unfolded proteins to cytosolic chaperonin. *Cell, 93*, 863–873.
23. Tian, G., Vainberg, I. E., Tap, W. D., Lewis, S. A., & Cowan, N. J. (1995). Quasi-native chaperonin-bound intermediates in facilitated protein folding. *The Journal of Biological Chemistry, 270*, 23910–23913.
24. Lopez-Fanarraga, M., Avila, J., Guasch, A., Coll, M., & Zabala, J. C. (2001). Review: Postchaperonin tubulin folding cofactors and their role in microtubule dynamics. *Journal of Structural Biology, 135*, 219–229.
25. Tian, G., Bhamidipati, A., Cowan, N. J., & Lewis, S. A. (1999). Tubulin folding cofactors as GTPase-activating proteins. GTP hydrolysis and the assembly of the alpha/beta-tubulin heterodimer. *Journal of Biological Chemistry, 274*, 24054–24058.
26. Nogales, E., Wolf, S. G., & Downing, K. H. (1998). Structure of the alpha beta tubulin dimer by electron crystallography. *Nature, 391*, 199–203.
27. Downing, K. H., & Nogales, E. (2010). Cryoelectron microscopy applications in the study of tubulin structure, microtubule architecture, dynamics and assemblies, and interaction of microtubules with motors. *Methods in Enzymology, 483*, 121–142.
28. Rajagopal, A., Kulkarni, S., Lewis, K. T., Chen, X., Maarouf, A., Kelly, C. V., Taatjes, D. J., & Jena, B. P. (2015). Proteome of the insulin-secreting Min6 porosome complex: Involvement of Hsp90 in its assembly and function. *Journal of Proteomics, 114*, 83–92.

29. Csermely, P. (2001). Chaperone overload is a possible contributor to 'civilization disease'. *Trends in Genetics, 17*, 701–704.
30. Slavotinek, A. M., & Biesecker, L. G. (2001). Unfolding the role of chaperones and chaperonins in human disease. *Trends in Genetics, 17*, 528–535.
31. Soti, C., & Csermely, P. (2002). Chaperones and aging: Role in neurodegeneration and in other civilizational diseases. *Neurochemistry International, 41*, 383–389.
32. Soti, C., & Csermely, P. (2003). Aging and molecular chaperones. *Experimental Gerontology, 38*, 1037–1040.
33. Macario, A. J. L., & Conway de Macario, E. (2002). Sick chaperones and ageing: A perspective. *Ageing Research Reviews, 1*, 295–311.
34. Macario, A. J. L., & Conway de Macario, E. (2000). Stress and molecular chaperones in disease. *International Journal of Clinical and Laboratory Research, 30*, 49–66.
35. Snoeckx, L. H. E. H., Cornelussen, R. N., van Nieuwenhoven, F. A., Reneman, R. S., & van der Vusse, G. J. (2001). Heat shock proteins and cardiovascular pathophysiology. *Physiological Reviews, 81*, 1461–1497.
36. Verbeke, P., Fonager, J., Clark, B. F. C., & Rattan, S. I. S. (2001). Heat shock response and ageing: Mechanisms and applications. *Cell Biology International, 25*, 845–857.
37. Hay, D. G., Sathasivam, K., Tobaben, S., Stahl, B., Marber, M., Mestril, R., Mahel, A., Smith, D. L., Woodman, B., & Bates, G. P. (2004). Progressive decrease in chaperone protein levels in a mouse model of Huntington's disease and induction of stress proteins as a therapeutic approach. *Human Molecular Genetics, 13*, 1389–1405.
38. Stone, D. L., Slavotinek, A., Bouffard, G. G., Banerjee-Basu, S., Baxevanis, A. D., Barr, M., & Biesecker, L. G. (2000). Mutations of a gene encoding a putative chaperonin causes McKusick-Kaufman syndrome. *Nature Genetics, 25*, 79–82.
39. Slavotinek, A. M., Dutra, A., Kpodzo, D., Pak, E., Nakane, T., Turner, J., et al. (2004). A female with complete lack of Müllerian fusion, postaxial polydactyly, and tetralogy of fallot: Genetic heterogeneity of McKusick-Kaufman syndrome or a unique syndrome? *American Journal of Medical Genetics, 129A*, 69–72.
40. Schwahn, U., Paland, N., Techritz, S., Lenzner, S., & Berger, W. (2001). Mutations in the X-linked RP2 gene cause intracellular misrouting and loss of the protein. *Human Molecular Genetics, 10*, 1177–1183.
41. Bartolini, F., Bhamidipatis, A., Thomas, S., Schwahn, U., Lewis, S. A., & Cowan, N. J. (2002). Functional overlap between retinitis pigmentosa 2 protein and the tubulin-specific chaperone cofactor C. *The Journal of Biological Chemistry, 277*, 14629–14634.
42. Stock, A. D., Spallone, P. A., Dennis, T. R., Netski, D., Morris, C. A., Mervis, C. B., & Hobart, H. H. (2003). Heat shock protein 27 gene: Chromosomal and molecular location and relationship to Williams syndrome. *American Journal of Medical Genetics, 120A*, 320–325.
43. Evgrafov, O. V., Mersiyanova, I., Irobi, J., van Den Bosch, L., Dierick, I., Leung, C. L., et al. (2004). Mutant small heat-shock protein 27 causes axonal Charcot-Marie-tooth disease and distal hereditary motor neuropathy. *Nature Genetics, 36*, 602–606.
44. Brady, J. P., Garland, D., Duglas-Tabor, Y., Robinson, W. G. J., Groome, A., & Wawrousek, E. F. (1997). Targeted disruption of the mouse alpha-A-crystallin gene induces cataract and cytoplasmic inclusion bodies containing the small heat shock protein alpha-B- crystallin. *Proceedings of the National Academy of Sciences of the United States of America, 94*, 884–889.
45. Litt, M., Kramer, P., La Morticella, D. M., Murphey, W., Lovrien, E. W., & Weleber, R. G. (1998). Autosomal dominant congenital cataract associated with a missense mutation in the human alpha crystallin gene CRYAA. *Human Molecular Genetics, 7*, 471–474.
46. Trent, J. M., Smith, L., & Brownstein, M. J. (1999). Progressive juvenile-onset punctate cataracts caused by mutations of the gammaD-crystallin gene. *Proceedings of the National Academy of Sciences of the United States of America, 96*, 1008–1012.

47. Sandilands, A., Hutcheson, A. M., Long, H. A., Prescott, A. R., Vrensen, G., Loster, J., et al. (2002). Altered aggregation properties of mutant gamma crystallins cause inherited cataract. *The EMBO Journal, 21*, 6005–6014.
48. Santhiya, S. T., Manohar, M. S., Rawlley, D., Vijayalakshmi, P., Namperumalsamy, P., Gopinath, P. M., Loster, J., & Graw, J. (2002). Novel mutation in the gamma-crystallin genes cause autosomal dominant congenital cataracts. *Journal of Medical Genetics, 39*, 352–358.

Chapter 4
ATP Synthase: Energy Generating Machinery in Cells

Abstract ATP synthase is an enzyme that drives the formation of ATP from ADP and Pi. It is a molecular rotary motor composed of F1 and Fo subunits and couples ATP synthesis during cellular respiration to an electrochemical gradient created by differences in proton concentration across mitochondrial membranes. In the mitochondria, proton is pumped across the inner mitochondrial membrane as electrons traveled through the electron transport chain. This results in a proton gradient, lowering the pH in the intermembrane space and elevated pH in the mitochondrial matrix. The proton gradient and membrane potential are the proton motive force that drives ATP synthesis.

Keywords Mitochondria · ATP synthesis · Rotary motor · Electrochemical gradient · Proton motive force · Electron transport chain

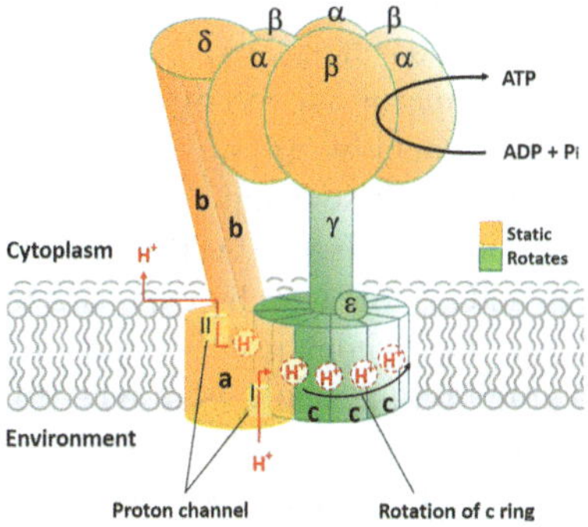

B. P. Jena, *Cellular Nanomachines*,
https://doi.org/10.1007/978-3-030-44496-9_4

Discovery and Structure–Function

Energy is required to sustain all life processes and forms. The energy currency for the unit of life the cell is adenosine triphosphate (ATP), which in eukaryotes is produced in the mitochondria. The discovery of the ATP generating machinery in cells was made in the 1950s and 1960s by the pioneering studies of *Efraim Racker* [1–4]. In the early 1950s, it was believed that ATP synthesis was coupled to respiration through a "High Energy" intermediate. To identify such an intermediate or enzyme involved in ATP synthesis (ATP synthase), mitochondrial membrane preparations from bovine heart which catalyzed oxidative phosphorylation were utilized. When mitochondrial fragments were sedimented in an ultracentrifuge, the fragments retained respiratory function, however, were unable to synthesize ATP due to the absence of oxidative phosphorylation [1–3]. Oxidative phosphorylation could be restored to the sedimented fragments on addition of the supernatant fraction. This first soluble protein fraction was coined F_1, required for coupling the phosphorylation of ADP to the oxidative electron transport chain to produce ATP. Subsequently, the membrane-associated mitochondrial particulate fraction was isolated using cholate and salt extraction [4]. This insoluble F_1-binding factor was named F_0 due to its sensitivity to oligomycin. Together, F_1F_0 therefore was demonstrated to constitute the ATP synthase holoenzyme. Then in 1961, *Peter D. Mitchell* made a major breakthrough in our understanding of ATP synthesis, when he proposed the chemiosmotic hypothesis [5]. The hypothesis states that electrons cross the membrane by linking to proton transport, generating proton motive force (PMF) to drive ATP synthesis [5]. PMF was shown to be established as a result of the differences in pH and electrical potential across the coupling membrane [6]. For example, in the mitochondria, proton would be pumped across the inner mitochondrial membrane as electrons traveled through the electron transport chain. This would result in a proton gradient, lowering the pH in the intermembrane space and elevated pH in the mitochondrial matrix. The proton gradient and membrane potential are the PMF that drives ATP synthesis. The pH gradient therefore acts like a battery, storing energy to produce ATP. Among the confirming experimental evidence in support of the chemiosmotic hypothesis is the elegant reconstitution experiment by *Efraim Racker* and *Walther Stoeckenius* [7]. In this study, a mitochondrial ATPase complex from beef heart reconstituted into an artificial lipid bilayer could synthesize ATP when the light-driven proton pump bacteriorhodopsin was added, followed by exposure to light [7]. It was next demonstrated by *Paul D. Boyer* that F_1 contains three catalytic domains [8] that store conformational energy required for the synthesis of ATP [9]. Rotary action involving the three catalytic domains was suggested [10, 11]; however, a major breakthrough in our understanding of how F_1 participates in ATP synthesis came in 1994 with an understanding of the crystal structure of F_1 [12, 13], from the laboratory of *John E. Walker*. In the 1994 study [12], the authors reported at 2.8 Å resolution the crystal structure of bovine mitochondrial F_1 ATPase. The study reported that the three catalytic beta subunits differ in conformation in the bound and unbound nucleotide states, and that the three catalytic subunits are in different states of the catalytic cycle at any given moment [12]. The crystal structure of F_1 provided the arrangement of its subunits and revealed its rotary property and catalytic activity (Fig. 4.1). The α- and β-subunits alternate and are arranged in a pseudo-hexagon motif, with the γ-subunit forming a central shaft. One nucleotide is bound to each of the three α-subunits. While one of the three β-subunit remains empty, a second binds to MgADP and a third binds to ATP. The γ-subunit central shaft serves as the crank, with its convex side pushing the helix-turn-helix motif at the C-terminus of the β-subunit. It is understood that binding of ATP to a β-subunit and its hydrolysis drives the crankshaft forward, leading to its interaction with the β-subunit hinge to push out the ADP and Pi following ATP hydrolysis. The direct observation of the rotation of F_1 ATPase was first made in 1997 in the laboratory of *Kazuhiko Kinosita Jr* [14]. In this study, a fluorescent actin filament serving as a marker attached to the γ-subunit enabled visualization of rotation in the presence of ATP [14]. More than 100 revolution of rotation in a counterclockwise direction was reported when viewed from the membrane side. Interestingly, a rotary torque of 40 nN nm^{-1} was observed in the study [14].

The F_0 is composed of three separate subunits with a stoichiometry of $C_{10}ab_2$. To conduct protons, all three subunits are required [15]. The c-subunit is highly hydrophobic and carries an ionizable residue imparted either by aspartic or glutamic acid that dictates the preference for H^+ or Na^+ binding to F_0 [16]. In the F_1F_0 holoenzyme, the two motors operate antagonistically. The net ATP turnover is zero when the thermodynamic driving forces are the same. In the presence of PMF, the F_1F_0 holoenzyme synthesizes ATP. However, if the molar free energy of ATP hydrolysis prevails, then the enzyme pumps protons and generates PMF. In the membrane of chloroplasts in photosynthetic organisms, the F_1F_0 holoenzyme is monomeric; however, in mammalian mitochondria, it is present as bands of dimers lining the mitochondria cristae rim. The mitochondria in mammalian cells hosts an F-ATPase called IF_1 [17] which blocks the rotation of the central shaft. During proton-driven ATP synthesis, the inhibitor is expelled; hence, the role of IF_1 in the mitochondria is

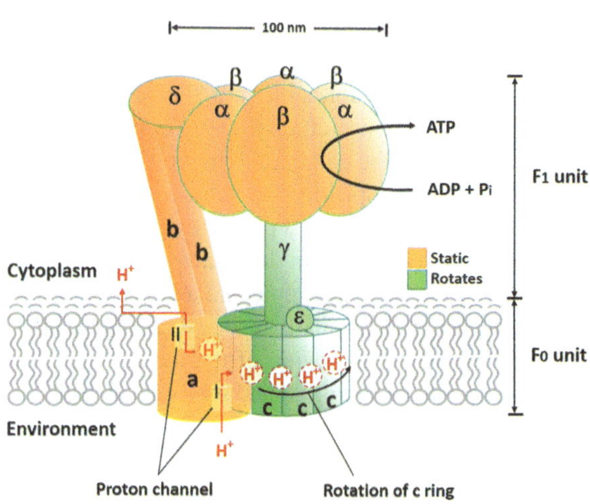

Fig. 4.1 Schematic diagram of the ATP synthase F_1F_0 holoenzyme nanomachine. Note in F_1 the alternating α- and β-subunits arranged in a pseudo-hexagon motif, and the γ-subunit forming a central shaft. The binding of ATP to a β-subunit and its hydrolysis drives the crankshaft forward. [*Modified from:* https://i.ytimg.com/vi/vIs8DOG5mZc/maxresdefault.jpg]

to prevent ATP hydrolysis under low PMF states [18]. Majority of cellular ATP in mammalian cells is generated via aerobic respiration in the mitochondria.

Disease Resulting from Altered ATP Synthase

As outlined earlier, the mitochondrial F_1F_0 ATP synthase is a 16-subunit complex with the globular catalytic F_1 portion connected by the stalks to the proton-translocating, membrane-spanning F_0 component. Two subunits of F_0 are encoded by the mitochondrial genome, i.e., MT-ATP6 and MT-ATP8 encoding subunits a and 8, whereas all the other ATP synthase subunits are encoded by nuclear genes. Defects in the structure and/or assembly of F_1F_0 ATP synthase are causal to a wide range of fatal diseases, primarily in children often shortly after birth. Mutations in both the nuclear and mitochondrial genes encoding F_1F_0 ATP synthase have been implicated in disease. For instance, nearly 60 different mutations in the mitochondria DNA genes MT-ATP6 and MT-ATP8 encoding subunits a and 8 of ATP synthase have been implicated in a wide range of diseases [19–21]. These include autism, a neurodevelopment disorder characterized by impaired energy metabolism, social interactions, and both verbal and non-verbal communications. Similarly, ataxia, encephalopathy, infantile cardiomyopathy, metabolic syndrome, motor neuron syndrome, lactic acidosis, schizophrenia, hereditary spastic paraplegia, and Charcot–Marie–Tooth syndrome are among some of the diseases resulting from ATP synthase subunit dysfunction. Similarly, ATP synthase enzyme deficiency due to nuclear genome mutations [22] results in impaired energy production, insufficient ATP synthase phosphorylating capacity, increase in mitochondria membrane potential and the consequent increase in production of reactive oxygen species (ROS) [23].

Each year, over ten million people are infected with the tuberculosis causing bacteria *Mycobacterium tuberculosis,* of which 1.5 million people die from the disease with half-million infected with multi-drug resistant bacteria [24–27]. Knowledge of the atomic structure of various components of ATP synthase is beginning to provide valuable clues for the development of novel drugs in combating pathogens such as bacterial disease resistance. A 2019 study [24], for example, demonstrates that the catalytic domain of ATP synthase in *Mycobacterium* is a target for the development of antitubercular drugs. The F_1 ATP synthase of *Mycobacterium tuberculosis* poorly hydrolyzes ATP. The crystal structure of F_1 ATP synthase of *Mycobacterium tuberculosis* suggests that its β_E-subunit is occupied by adenosine 5′-diphosphate molecule and no magnesium ion and phosphate, exhibiting arrest conformation in their rotary catalytic cycle. Screening of small molecules that could bind to this region of the motor would help abrogate ATP synthesis by *Mycobacterium tuberculosis* and inhibit its growth, without impacting the host ATP synthase function.

References

1. Pullman, M. E., Penefsky, H., & Racker, E. (1958). A soluble protein fraction required for coupling phosphorylation to oxidation in submitochondrial fragments of beef heart mitochondria. *Archives of Biochemistry and Biophysics, 76*, 227–230.
2. Pullman, M. E., Penefsky, H. S., Datta, A., & Racker, E. (1960). Partial resolution of the enzymes catalyzing oxidative phosphorylation. I. Purification and properties of soluble, dinitrophenol-stimulated adenosine triphosphatase. *The Journal of Biological Chemistry, 235*, 3322–3329.
3. Penefsky, H. S., Pullman, M. E., Datta, A., & Racker, E. (1960). Partial resolution of the enzymes catalyzing oxidative phosphorylation II. Participation of a soluble adenosine triphosphatase in oxidative phosphorylation. *Journal of Biological Chemistry, 235*, 3330–3336.
4. Kagawa, Y., & Racker, E. (1966). Partial resolution of the enzymes catalyzing oxidative phosphorylation IX. Reconstruction of oligomycin-sensitive adenosine triphosphatase. *Journal of Biological Chemistry, 241*, 2467–2474.
5. Mitchell, P. (1961). Coupling of photophosphorylation to electron and hydrogen transfer by a chemiosmotic type of mechanism. *Nature, 191*, 144–148.
6. Mitchell, P. (1966). Chemiosmotic coupling in oxidative and photosynthetic phosphorylation. *Physiological Reviews, 41*, 445–502.
7. Racker, E., & Stoeckenius, W. (1974). Reconstitution of purple membrane vesicles catalyzing light-driven proton uptake and adenosine triphosphate formation. *The Journal of Biological Chemistry, 249*, 662–663.
8. Kayalar, C., Rosing, J., & Boyer, P. D. (1977). An alternating site sequence for oxidative phosphorylation suggested by measurement of substrate binding patterns and exchange reaction inhibitions. *The Journal of Biological Chemistry, 252*, 2486–2491.
9. Boyer, P. D. (1977). Conformational coupling in oxidative phosphorylation and photophosphorylation. *Trends in Biochemical Sciences, 2*, 38–41.
10. Boyer, P. D., & Kohlbrenner, W. E. (1981). The present status of the binding change mechanism and its relation to ATP formation by chloroplasts. In B. R. Selman & S. Selman-Reimer (Eds.), *Energy coupling in photosynthesis* (pp. 231–241). Amsterdam: Elsevier.
11. Rao, R., & Senior, A. E. (1987). The properties of hybrid F1-ATPase enzymes suggest that a cyclical catalytic mechanism involving three catalytic sites occurs. *The Journal of Biological Chemistry, 25*, 17450–17454.
12. Abrahams, J. P., Leslie, A. G., Lutter, R., & Walker, J. E. (1994). The structure of F1-ATPase from bovine heart mitochondria determined at 2.8 a resolution. *Nature, 370*, 621–628.
13. Bowler, M. W., Montgomery, M. G., Leslie, A. G., & Walker, J. E. (2007). Ground state structure of F1-ATPase from bovine heart mitochondria at 1.9 a resolution. *The Journal of Biological Chemistry, 282*, 14238–14242.
14. Noji, H., Yasuda, R., Yoshida, M., & Konisita, K., Jr. (1997). Direct observation of the rotation of F1-ATPase. *Nature, 386*, 299–302.
15. Schneider, E., & Altendorf, K. (1985). All three subunits are required for the reconstitution of an active proton channel (FO) of *Escherichia coli* ATP synthase (F1FO). *The EMBO Journal, 4*, 515–518.
16. Preiss, L., Yildiz, O., Hicks, D. B., Krulwich, T. A., & Meier, T. (2010). A new type of proton coordination in an F1FO-ATP synthase rotor ring. *PLoS Biology, 8*, e1000443.
17. Runswick, M. J., Bason, J. V., Montgomery, M. G., Robinson, G. C., Fearnley, I. M., & Walker, J. E. (2013). The affinity purification and characterization of ATP synthase complexes from mitochondria. *Open Biology, 3*, 120160.
18. Bason, J. V., Montgomery, M. G., Leslie, A. G., & Walker, J. E. (2014). Pathway of binding of the intrinsically disordered mitochondrial inhibitor protein to F1-ATPase. *Proceedings of the National Academy of Sciences, 111*, 11305–11310.
19. Dimauro, S., & Schon, E. A. (2003). Mitochondrial respiratory-chain diseases. *The New England Journal of Medicine, 348*, 2656–2668.

20. Zeviani, M., & Carelli, V. (2007). Mitochondrial disorders. *Current Opinion in Neurology, 20*, 564–571.
21. Vafai, S. B., & Mootha, V. K. (2012). Mitochondrial disorders as windows into an ancient organelle. *Nature, 491*, 374–383.
22. Housˇteˇk, J., Klement, P., Floryk, D., Antonicka, H., Hermanska, J., Kalous, M., Hansikova, H., Hout'kova, H., Chowdhury, S. K., Rosipal, T., et al. (1999). A novel deficiency of mitochondrial ATPase of nuclear origin. *Human Molecular Genetics, 8*, 1967–1974.
23. Mracek, T., Pecina, P., Vojtiskova, A., Kalous, M., Sebesta, O., & Housˇteˇk, J. (2006). Two components in pathogenic mechanism of mitochondrial ATPase deficiency: Energy deprivation and ROS production. *Experimental Gerontology, 41*, 683–687.
24. Zhang, A. T., Montgomery, M. G., Leslie, A. G. W., Cook, G. M., & Walker, J. E. (2019). The structure of the catalytic domain of the ATP synthase from Mycobacterium smegmatis is a target for developing antitubercular drugs. *Proceedings of the National Academy of Sciences of the United States of America, 116*, 4206–4211.
25. World Health Organization. (2018). Global Tuberculosis Report 2018. Retrieved September 18, 2018 from https://www.who.int/tb/publications/global_report/en/
26. Houben, R. M., & Dodd, P. J. (2016). The global burden of latent tuberculosis infection: A re-estimation using mathematical modelling. *PLoS Medicine, 13*, e1002152.
27. Behr, M. A., Edelstein, P. H., & Ramakrishnan, L. (2018). Revisiting the timetable of tuberculosis. *BMJ, 362*, k2738.

Chapter 5
Ribosome: Cells Protein Synthetic Machinery

Abstract Ribosome is the translational apparatus in cells, serving as the protein synthetic machinery. Specified by mRNA, ribosomes link amino acids to form polypeptides. Ribosomes are composed of two subunits: a large and a small subunit. The small subunit reads the mRNA and the large subunit links amino acids to form polypeptides. Each subunit is composed of ribosomal RNA and ribosomal protein molecules; hence, it is a ribonucleoprotein. In protein synthesis, the small ribosomal subunit bound to an aminoacyl-tRNA binds to the start codon on the mRNA and recruits the large ribosomal subunit. At the inter-subunit region, ribosomes have three transfer RNA or tRNA binding sites: the A or aminoacyl site, the P or peptidyl site, and the E for the exit site. At the 30S small ribosome subunit, the mRNA binds to a track, moving in a stepwise manner, one codon at a time, resulting in peptide elongation and termination following completion of protein synthesis.

Keywords Translational apparatus · Ribosomal RNA · Ribosomal proteins

Discovery

In cells, genetic information encoded in the DNA is transcribed into mRNA by RNA polymerase. The resulting mRNA is then translated into proteins from amino acids by 25–30 nm ribosomes associated with the membrane of the endoplasmic reticulum. Hence, ribosomes are the protein synthetic nanomachine in cells. In 1937, studies by *Albert Claude* involving differential centrifugation identified in Rous sarcoma tumor-induced fractions enriched the presence of ribonucleic acid [1]. This enriched ribonucleic acid fraction was inactivated by agents then known to affect nucleoproteins and nucleic acids [2]. Chick embryo fractions [3] and fractions from a wide range of adult and embryonic tissue were also found to possess similar ribonucleic acid-enriched fractions [4], suggesting their universal presence

B. P. Jena, *Cellular Nanomachines*,
https://doi.org/10.1007/978-3-030-44496-9_5

in normal cells. Using cell fractionation and centrifugation studies, *Albert Claude* arrived at the conclusion that the ribonucleic acid-enriched fractions are particulate measuring 50–200 nm and termed them "microsomes" [5]. From the 1930s to the early 1950s, although much progress was made on the chemical composition of microsomes, little progress was made on their morphology and cellular distribution, primarily due to limitation in resolution of the light microscope. The invention of the electron microscope (EM) and the development of appropriate procedures for biological sample preparation for EM finally led to discovery and the physical presence and distribution of the microsome in cells. Microsomes were first observed in the 1950s by *George Emil Palade* using the transmission EM [6] as tiny granular objects either associated with the endoplasmic reticulum (ER) and/or dispersed in the cytosol [6]. The "microsomes" or "microsomal particles" were later demonstrated to be composed of both RNA and proteins and to be involved in protein synthesis [7]. To overcome confusion with other subcellular structures and the particles association with the ER membrane, at the 1958 Biophysical Society Meetings, the term "ribosome" was suggested and adopted [8]. Immediately thereafter, although the *"one gene one ribosome"* was first proposed followed by "ribosomal RNA act as template" for protein synthesis [9], studies on the kinetics of β-galactosidase synthesis supported that ribosomes *"make one protein at one moment and a different protein later"* [10]. In an elegant study [11] involving bacteriophage infection of *E. coli*, although new RNA corresponding to the phage genome was observed to be synthesized, no new ribosomes were made. These results demonstrated for the first time that *"ribosomes are non-specialized structures which synthesize at a given time the protein dictated by the messenger (mRNA) they happen to be associated with"* [11]. Recent studies, however, demonstrate the existence of heterogeneity in ribosome composition, imparting them their additional specialized functions [12, 13].

Ribosomes catalyze two key chemical reactions, one involving peptide bond formation for chain elongation and the other involves ester bond hydrolysis for chain termination [14–17]. The molecular mechanism of these chemical reactions remained a mystery for decades, since no high resolution structures were available. The major hurdle was in the large size (2.4 MDa) of the ribosome, its asymmetry and phase problems, rendering it difficult to obtain appropriate crystals for X-ray crystallography. The major breakthrough came with the pioneering structural studies by *Ada E. Yonath* through her development and use of the new approach of cryo-crystallography [18]. All crystallization was carried out at 4 °C–7 °C using 30S and 50S ribosomal subunits isolated from *Bacillus stearothermophilus* [18]. The generated crystals were stable at 6 °C even when exposed to X-rays up to 30 h [18]; however, when transferred to room temperature, the crystals rapidly dissolved, thus began the new approach of cryo-crystallography. Once ribosome crystals were available, *Ada E. Yonath, Thomas A. Steitz, Venkatraman Ramakrishnan*, and other groups were able to solve the bacterial 50S, 30S, and the entire (70S) ribosome structure at near atomic resolution utilizing X-ray crystallography and cryo-electron microscopy [19–28], and in doing so, unraveled the molecular mechanism of its function.

To solve the ribosome structure at atomic resolution, phases for the different X-ray diffraction patterns had to be determined for this large uneven structure. Utilizing cryo-electron microscopy [21] combined with multiple isomorph replacement and scattering approaches, *Joachim Frank* helped to initially resolve this issue, leading to the low resolution X-ray structures of the 50S subunit by *Thomas A. Steitz* [22, 23]. The final breakthrough in the atomic structure of the ribosome came in 2000 by *Thomas A. Steitz* for the 50S subunit [24] and by *Venkatraman Ramakrishnan* for the 30S subunit [25]. The atomic structure of the two ribosomal subunits provided new insights into the structure and additional high resolution X-ray studies revealed secrets of the function of this protein synthetic nanomachine in cells. Fundamental understanding of how this nanomachine accurately determines tRNA selection and peptidyl transfer activity during peptide elongation was made.

Structure–Function

In bacteria, ribosomes (70S) are composed of a small 30S and a large 50S subunit, of molecular mass 0.8 MkDa and 1.5 MkDa, respectively (Fig. 5.1). The 30S subunit is further composed of 20 different proteins and 16S ribosomal RNA or rRNA, containing nearly 1600 nucleotides. Similarly, the 50S large ribosomal subunit is made up of 33 different proteins and a 23S and 5S rRNA with 2900 and 120 nucleotides, respectively. Although ribosomes in eukaryotes are larger and more complex

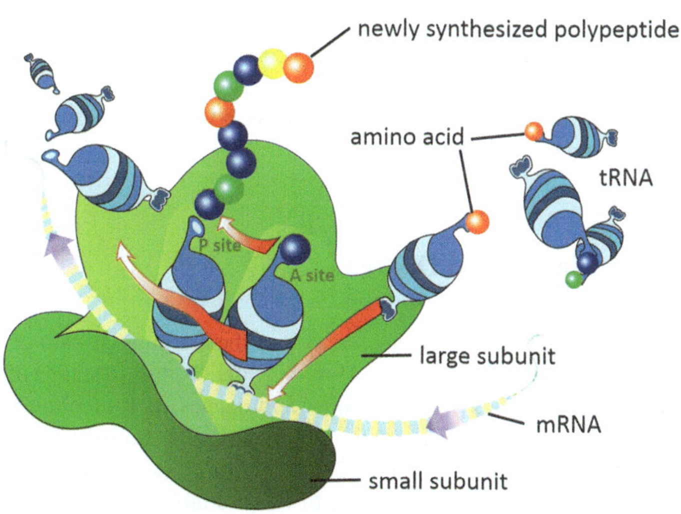

Fig. 5.1 Schematic drawing of the ribosome the protein synthetic machinery in cells and associated molecules. The large and small ribosomal subunits of the machinery, and associated mRNA, amino acid-tRNA complexes, and newly synthesized polypeptide, are depicted. [*Modified version generated from:* https://en.wikipedia.org/wiki/Translation_(biology)]

than ribosomes in prokaryotes, all ribosomes exhibit similar general structure and functions (Fig. 5.1). At the inter-subunit region, ribosomes have three transfer RNA or tRNA binding sites, namely the A or aminoacyl site, the P or peptidyl site, and the E for the exit site. At the 30S small ribosome subunit, the mRNA binds to a track, moving in a stepwise manner, one codon at a time, resulting in peptide elongation and protein synthesis. Protein synthesis is initiated when the leader sequence binds with the anti-leader sequence of the 16S RNA. This defined chain of events is followed by the binding of the initiator tRNA (associated with formylated methionine) to the P site, a reaction which is accelerated by the three initiation factors IF1, IF2 (GTPase), and IF3. The proper configuration (correct reading frame) of mRNA and initiator tRNA promotes the docking of the 50S ribosome large subunit, resulting in the assembly of the 70S peptide elongation phase of the ribosome complex. To promote elongation, the aminoacyl-tRNA complexed with the elongation factor Tu (EF-Tu) which is a GTPase, enters the A site. The standard free energy states of codon–anticodon pairing dictate the precise translation of the genetic code to amino acid sequence through aminoacyl-tRNA cognate to mRNA codons displayed at the A site. In response to appropriate codon–anticodon interaction, there is hydrolysis of GTP by EF-Tu, enabling aminoacyl-tRNA binding to the A site and peptide bond formation catalyzed at the peptidyl-transfer center of the 50S ribosome subunit [14]. Next, elongation factor G (EF-G), also a GTPase, binds and translocates peptidyl-tRNA from the A to the P site while forwarding the mRNA in its reading frame. This process enables the A site to associate with the next codon to be read by aminoacyl-tRNA [15]. In this way, peptide elongation continues until a stop codon appears in the A site in protein synthesis. Stop codons are read by release factors 1 (RF1, codons UAA, UAG) and release factors 2 (RF2, codons UAA, UGA) that induce hydrolysis of the ester bond linking a completed peptide chain with the P site bound tRNA. This leads to the release of the synthesized peptide to undergo proper folding and the dissociation of the release factors by factor 3 (RF3, a GTPase). The ribosome can then undergo a new round of protein synthesis by a new mRNA [16, 17], thus completing a cycle.

Ribosome in Health and Disease

In the past decade, there is increasing concern on the alarming growth of antibiotic resistance among bacterial pathogens. However, the atomic structure of the ribosome now available [18–25] provides new approaches in overcoming antibiotic resistance. It comes as no surprise that the ribosome has become the target of nearly half of all antibacterial drugs currently available. In addition to various components of the ribosome structure, a large number of the available antibiotic drugs bind to the peptidyl-transfer center of the 50S ribosome subunit to overcome antibiotic resistance [26–28].

A large number of clinical phenotypes resulting from impaired biogenesis of ribosomes have been reported. Mutation in a ribosomal protein gene RPS19 is reported in patients with Diamond–Blackfan anemia (DBA), a rare bone marrow disease with erythroid defects [29]. Similarly, defective ribosome biogenesis has been linked to a wide range of congenital diseases such as Shwachman–Diamond syndrome (SDS), cartilage hair hypoplasia, Treacher-Collins syndrome (TCS), X-linked dyskeratosis congenita, and 5q-syndrome [30, 31]. DBA is characterized by a decrease in erythroid precursors, anemia, macrocytosis, and reticulocytopenia. Majority of patients are children diagnosed in their first year of life, with pallor and lethargy as presenting symptoms. DBA patients exhibit higher levels of adenosine deaminase in their red blood cells, possess short stature, and frequently exhibit various cardiac defects. Current definitive therapy for DBA patients is limited to bone marrow transplant. Similarly, SDS patients exhibit mutation in the highly conserved SBDS gene that plays a role in ribosome biogenesis and processing of both rRNA and mRNA [32, 33]. TCS similarly has been identified to mutation in the TCOF1 gene [34] that encodes the treacle protein, a part of the pre-ribosomal ribonucleoprotein complex. Treacle is required for the transcription of the ribosomal DNA and may also play a role in rRNA methylation [35]. Patients with TCS often exhibit complications from craniofacial dysostosis, with complications in their airways, and exhibit defects in brain development and hearing phenotypes. Interestingly, studies in mice demonstrate that the inhibition of p53 activity rescues craniofacial abnormalities as a consequence of haplo-insufficient TCOF1 gene that demonstrates diminished production of ribosomes [36, 37].

References

1. Claude, A. (1937). Properties of the causative agent of a chicken tumor: XIII. *The Journal of Experimental Medicine, 66*, 59–72.
2. Claude, A., & Rothen, A. (1940). Properties of the causative agent of a chicken tumor: XIV. *The Journal of Experimental Medicine, 71*, 619–633.
3. Claude, A. (1938). A fraction from normal chicken embryos similar to the tumor producing fraction of chicken tumors. *Proceedings of the Society for Experimental Biology and Medicine, 39*, 398.
4. Claude, A. (1940). Particulate component of normal and tumor cells. *Science, 91*, 77.
5. Claude, A. (1941). Particulate component of cytoplasm. *Cold Spring Harbor Symposia on Quantitative Biology, 9*, 263.
6. Palade, G. E. (1955). A small particulate component of the cytoplasm. *The Journal of Cell Biology, 1*, 59–68.
7. Palade, G. E. (1958). Microsomes and ribonucleoprotein particles. In R. B. Roberts (Ed.), *Microsomal Particles and Protein Synthesis* (pp. 36–61). London: Pergamon Press.
8. Roberts, R. B. (1958). Introduction. In *Microsomal Particles and Protein Synthesis* (pp. vii–viii). London: Pergamon Press.
9. Crick, F. H. C., & Brenner, S. (1959). *Some footnotes on protein synthesis: Note for the RNA TIE club*.
10. Brenner, S., & Crick, F. H. C. (1960). What are the properties of genetic RNA? https://profiles.nlm.nih.gov/ps/access/SCBBFZ.pdf.
11. Brenner, S., Jacob, F., & Meselson, M. (1961). An unstable intermediate carrying information from genes to ribosomes for protein synthesis. *Nature, 190*, 576–581.

12. Shi, Z., Fujii, K., Kovary, K. M., Genuth, N. R., Rost, H. L., Teruel, M. N., & Barna, M. (2017). Heterogeneous ribosomes preferentially translate distinct subpools of mRNAs genome-wide. *Molecular Cell, 67*, 71–83.
13. Simsek, D., Tiu, G. C., Flynn, R. A., Byeon, G. W., Xu, A. F., Chang, H. Y., & Barnam, M. (2017). The mammalian ribo-interactome reveals ribosome functional diversity and heterogeneity. *Cell, 169*, 1051–1065.
14. Rodnina, M. V., Beringer, M., & Wintermeyer, W. (2007). How ribosomes make peptide bonds. *Trends in Biochemical Sciences, 32*, 20–26.
15. Frank, J., Gao, H., Sengupta, J., Gao, N., & Taylor, D. J. (2007). The process of mRNA-tRNA translocation. *Proceedings of the National Academy of Sciences, 104*, 19671–19678.
16. Karimi, R., Pavlov, M. Y., Buckingham, R. H., & Ehrenberg, M. (1999). Novel roles for classical factors at the interface between translation termination and initiation. *Molecular Cell, 3*, 601–609.
17. Zavialov, A. V., Mora, L., Buckingham, R. H., & Ehrenberg, M. (2002). Release of peptide promoted by the GGQ motif of class 1 release factors regulates the GTPase activity of RF3. *Molecular Cell, 10*, 789–798.
18. Yonath, A., Mussig, J., Tesche, B., Lorenz, S., Erdmann, V. A., & Wittmann, H. G. (1980). Crystallization of the large ribosomal subunits from Bacillus stearothermophilus. *Biochemistry International, 1*, 428–435.
19. Yonath, A., Bartunik, H. D., Bartels, K. S., & Wittmann, H. G. (1984). Some x-ray diffraction patterns from single crystals of the large ribosomal subunit from Bacillus stearothermophilus. *Journal of Molecular Biology, 177*, 201–206.
20. Shevack, A., Gewitz, H. S., Hennemann, B., Yonath, A., & Wittmann, H. G. (1985). Characterization and crystallization of ribosomal particles from *Halobacterium marismortui*. *FEBS Letters, 184*, 68–71.
21. Frank, J., Zhu, J., Penczek, P., Li, Y., Srivastava, S., Verschoor, A., Radermacher, M., Grassucci, R., Lata, R. K., & Agrawal, R. K. (1995). A model of protein synthesis based on cryo-electron microscopy of the E. coli ribosome. *Nature, 376*, 441–444.
22. Ban, N., Freeborn, B., Nissen, P., Penczek, P., Grassucci, R. A., Sweet, R., Frank, J., Moore, P. B., & Steitz, T. A. (1998). A 9 a resolution X-ray crystallographic map of the large ribosomal subunit. *Cell, 93*, 1105–1115.
23. Ban, N., Nissen, P., Hansen, J., Capel, M., Moore, P. B., & Steitz, T. A. (1999). Placement of protein and RNA structures into a 5 A-resolution map of the 50S ribosomal subunit. *Nature, 400*, 841–847.
24. Ban, N., Nissen, P., Hansen, J., Moore, P. B., & Steitz, T. A. (2000). The complete atomic structure of the large ribosomal subunit at 2.4 A resolution. *Science, 289*, 905–920.
25. Wimberly, B. T., Brodersen, D. E., Clemons, W. M., Jr., Morgan-Warren, R. J., Carter, A. P., Vonrhein, C., Hartsch, T., & Ramakrishnan, V. (2000). Structure of the 30S ribosomal subunit. *Nature, 407*, 327–339.
26. Franceschi, F., & Duffy, E. M. (2006). Structure-based drug design meets the ribosome. *Biochemical Pharmacology, 71*, 1016–1025.
27. Brodersen, D. E., Clemons, W. M., Jr., Carter, A. P., Morgan-Warren, R. J., Wimberly, B. T., & Ramakrishnan, V. (2000). The structural basis for the action of the antibiotic tetracycline, pactamycin, and hygromycin B on the 30S ribosomal subunit. *Cell, 103*, 1143–1154.
28. Carter, A. P., Clemons, W. M., Brodersen, D. E., Morgan-Warren, R. J., Wimberly, B. T., & Ramakrishnan, V. (2000). Functional insights from the structure of the 30S ribosomal subunit and its interactions with antibiotics. *Nature, 407*, 340–348.
29. Draptchinskaia, N., Gustavsson, P., Andersson, B., et al. (1999). The gene encoding ribosomal protein S19 is mutated in Diamond-Blackfan anaemia. *Nature Genetics, 21*(2), 169–175.
30. Liu, J. M., & Ellis, S. R. (2006). Ribosomes and marrow failure: Coincidental association or molecular paradigm? *Blood, 107*(12), 4583–4588.
31. Ebert, B. L., Pretz, J., Bosco, J., et al. (2008). Identification of RPS14 as a 5q− syndrome gene by RNA interference screen. *Nature, 451*(7176), 335–339.

32. Ganapathi, K. A., Austin, K. M., Lee, C. S., et al. (2007). The human Shwachman-Diamond syndrome protein, SBDS, associates with ribosomal RNA. *Blood, 110*(5), 1458–1465.
33. Rujkijyanont, P., Adams, S. L., Beyene, J., & Dror, Y. (2009). Bone marrow cells from patients with Shwachman-Diamond syndrome abnormally express genes involved in ribosome biogenesis and RNA processing. *British Journal of Haematology, 145*(6), 806–815.
34. Treacher Collins Syndrome Collaborative Group. (1996). Positional cloning of a gene involved in the pathogenesis of Treacher Collins syndrome. *Nature Genetics, 12*(2), 130–136.
35. Valdez, B. C., Henning, D., So, R. B., Dixon, J., & Dixon, M. J. (2004). The Treacher Collins syndrome (TCOF1) gene product is involved in ribosomal DNA gene transcription by interacting with upstream binding factor. *Proceedings of the National Academy of Sciences of the United States of America, 101*(29), 10709–10714.
36. Dixon, J., Jones, N. C., Sandell, L. L., et al. (2006). Tcof1/Treacle is required for neural crest cell formation and proliferation deficiencies that cause craniofacial abnormalities. *Proceedings of the National Academy of Sciences of the United States of America, 103*(36), 13403–13408.
37. Jones, N. C., Lynn, M. L., Gaudenz, K., et al. (2008). Prevention of the neurocristopathy Treacher Collins syndrome through inhibition of p53 function. *Nature Medicine, 14*(2), 125–133.

Chapter 6
Nuclear Pore: A Bidirectional Transport Machinery

Abstract The human nuclear pore complex is a 120 nm structure composed of nearly 1000 protein molecules (multiple copies of nearly 30 different proteins) with a mass of 110–120 MDa. It spans the double membrane of the nuclear envelope and selectively transports both proteins, nucleic acids, and small signaling molecules bidirectionally. The diameter of the channel in the nuclear pore complex is approximately 5 nm in diameter and 45 nm in depth. Selective transport through the nuclear pore complex is mediated by nuclear transport receptors that bind to the cargo to be transported. Importins mediate transport of cargo molecules into the nucleus, whereas exportins facilitate the selective transport of cargo out of the nucleus. Cargoes with a nucleus localization signal are efficiently transported into the nucleus through the nuclear pore complex. The import and export cycles require GTP hydrolysis, and thus the transport process through the nuclear pore complex is energy-dependent. Since the nuclear pore complex is the gateway to the genome, the number of nuclear pore complexes varies during the different stages of the cell cycle. For example, between G1 and G2 phase of the cell cycle, the number of nuclear pore complexes at the nuclear envelope increase to accommodate greater transcriptional demand. Assembly of the nuclear pore complex like other cellular nanomachines is little understood.

Keywords Nuclear envelope · Selective bidirectional pore · Nucleoporins

B. P. Jena, *Cellular Nanomachines*,
https://doi.org/10.1007/978-3-030-44496-9_6

Discovery

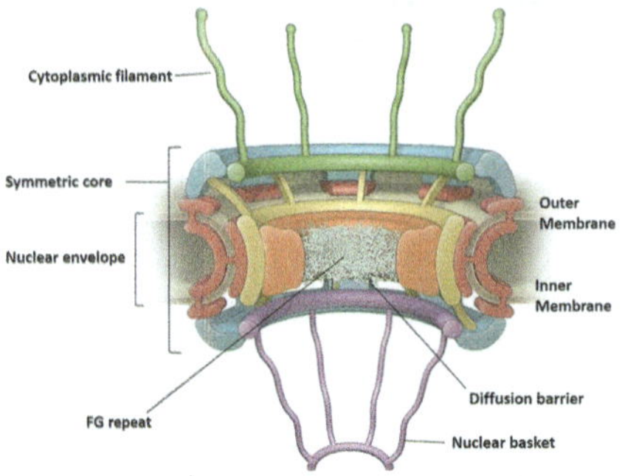

The nuclear pore complex was discovered in 1949 using electron microscopy (EM) [1]. Subsequently, detailed EM studies on the nuclear envelope of frog oocyte demonstrated the presence of pores in the double bilayer membrane of the nucleus [2]. In 1959, the protein complex embedded in the nuclear pore was first described [3]. This 1959 study [3] reported *"channels associated with the pore complex extending deep into the nucleus."* The study [3] further reported observation of crescents and spirals of ribonucleoprotein particles in the immediate vicinity of the outer nuclear pore membrane. The next breakthrough came in 1982, with the isolation and identification of the first major nuclear pore protein [4]. In this study [4], rat liver nuclear envelop was fractionated, and a major nuclear envelope glycoprotein gp190 was identified and a monospecific polyclonal antibody was raised for immunoisolation of the pore complex and for immunolocalization of the protein within the complex. In the past two decades, in addition to advances in nuclear pore morphology, 34 different nucleoporins have been identified [5]. Varying number of copies of each of these 34 nucleoporin molecules (Table 6.1), amounting to nearly 1000 molecules with a collective molecular mass of 120 MDa, constitute each of the 120 nm nuclear pore complexes in eukaryotes [6–12].

Structure–Function

The next breakthrough in our understanding of the nuclear pore came with its reconstitution [13] and the subsequent reconstitution studies on its assembly and function [14–17]. The *Xenopus* egg played a central role in our understanding of the nuclear pore assembly and function. The ability of extracts from an externally fertilizable *Xenopus* egg to reconstitute nuclei [13] was due to the fact that all components required for multiple rounds of rapid division and growth were present in the egg. Hence, irrespective of the DNA or chromatin source, their mere addition to *Xenopus* egg extracts in a tube sets in motion the assembly of hundreds of nuclei with their own nuclear pore complexes. This reconstitution provided an understanding of the

Table 6.1 List of mammalian nucleoporins, their yeast homologues, and their relative abundance (*Taken from Cronshaw* et al., *2002, ref.* 8)

Nucleoporin	Relative abundance	Yeast homologue(s)	Relative abundance
Nup358	8	–	–
Tpr	16	Mlp1p, Mlp2p	ND
Nup214	8	Nup159p	8
gp210	16	–	–
Nup205	16	Nup192p	16
Nup153	8	Nup1p[c]	8
Nup188	8	Nup188p	16
POM121	8	–	–
Nup155	32	Nup157p, Nup170p	32
Nup160	8	Nup120p	16
Nup133	16	Nup133p	16
Nup96	16	C-Nup145p	16
Nup107	32	Nup84p	16
Nup98	8	N-Nup145p, Nup116p, Nup100p	32
Nup93	32–48	Nic96p	32
Nup88	32	Nup82p	8–16
Nup62	16	Nsp1p	32
Nup75 (FLJ12549)	16	Nup85p	16
Nup58	48	Nup49p	16
ALADIN	8	–	–
Nup54	32–48	Nup57p	16
Nup50	32	Nup2p[c]	ND
Nup45	32	(Nup49p)	(16)
NLP1	16	Nup42p	8
Nup43 (p42)	16	–	–
RAE1	48	Gle2p	16–32
Seh1 (Sec13L)	16–32	Seh1p	16
Nup37 (p37)	16–32	–	–
Sec13R	16–32	Sec13p	ND
Nup35 (MP-44)	16–32	Nup53p, Nup59p	32

sequential order of the assembly of the nuclear pore complex, in addition to the arrangement of each sub-structure and the critical players involved in each stage of the pore assembly process. Furthermore, using different fractions of the extract has enabled the partial assembly of the nuclear pore complex, its function (cargo transport), and the identification of the players involved. In vitro assays have also allowed the use of membrane-free pore intermediate structures to study nuclear pore assembly, structure, and function.

Electron and atomic force microscopy of the nuclear pore complex demonstrate an eightfold rotational symmetry. Additionally, the nuclear and cytoplasmic halves of the complex are identical; however, they are rotated 180° relative to each other. The interior core is termed the symmetry core, composed of three ring structures (Fig. 6.1): an inner ring that surrounds the central transport channel and two outer rings flanking either sides of the nuclear envelope. Nuclear porins specific to the nuclear compartment associate with the symmetric core to form a nuclear basket. The nuclear basket interacts with chromatin and the transcription machinery. On the cytoplasmic compartment of the nuclear pore complex, nuclear porins form flexible cytoplasmic filaments that facilitate transport through the nuclear pore complex. The general architecture of the nuclear pore complex is highly conserved from humans to yeast. However, a major difference between them is size and mass, dictated by the presence of different number of copies of the various nucleoporin molecules. While the human nuclear pore complex has a mass of 110–120 MDa, the nuclear pore complex in yeast is nearly half that, with a mass of approximately 52 MDa. In contrast to channels and transporters, the nuclear pore does not open or close, rather it helps in the selective and active diffusion of molecules. While molecules as small as 3 nm are prevented from entry and transport through the diffusion barrier, selective cargo molecules of up to 40 nm in size are able to pass through.

The general organization of the diffusion barrier in the nuclear pore complex is poorly understood and continues to be a challenge. However, the organization of the evenly spaced nucleoporins lining the central transport channel has shed some light on the structure of the diffusion barrier. The diffusion barrier is formed by a large number of nucleoporins that contain stretches of phenylalanine–glycine (FG) repeats. Projection of the FG repeats into the transport channel creates a mesh-like selective filter. Although progress has been made in elucidating the molecular architecture of a number of nucleoporins and their interactions with other nucleoporins at certain domains of the nuclear pore complex, *how individual domains assemble and integrate to form the complex is far from clear*. Folded proteins in their native

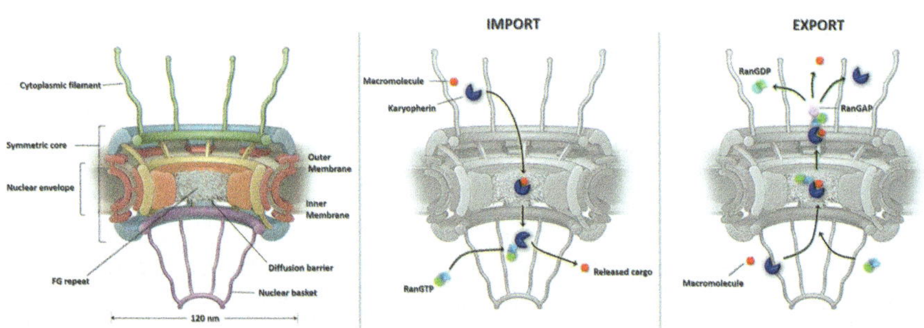

Fig. 6.1 Schematic drawing of the nuclear pore complex showing bidirectional transport (import and export). Some major components of the nuclear pore complex such as the cytoplasmic filament, the phenylalaning glycine (FG) repeats at the pore, and the nuclear basket facing the interior of the nucleus are shown. [*Modified version generated from:* https://www.the-scientist.com/infographics/infographic-the-nuclear-pore-complex-32456]

state pass through the diffusion barrier with help from molecules called karyopherins. There are 19 known members of karyopherins in humans, each specialized to recognize its own cargo and specialized in either import or export of molecules through the bidirectional nuclear pore complex.

An important role of the cytoplasmic filament of the nuclear pore complex is the export of mRNA. This is achieved by the spatially restricted activation of the ATPase DDX19 and through the diffusion barrier by heterodimeric transport factor Nxf1-Nxt1. DDX19 with a N-terminus β-propeller domain is a member of the protein family referred to as DEAD-box RNA helicase. The ATPase activity of DDX19 facilitates release of mRNAs from Nxf1-Nxt1 into the cytoplasm. However, how the ATPase activity of DDX19 is coupled to mRNA remodeling and release remains to be understood.

The nuclear basket nucleoporins decorate the nuclear pore facing the nuclear compartment, imparting mRNA anchoring site for export and to interact with the nucleocytoplasmic transport process and the transcription regulation machine. Ran-binding domains in the nuclear basket nucleoporin serve for the binding of the small GTPase Ran, which may participate as a disassembly factor for the nuclear export complex. Direct interaction between nuclear basket nucleoporins and genomic loci has been reported to help in coordinating transcription and mRNA processing and export by the nuclear pore complex.

Defects in Nuclear Transport and Disease

As the conduit for the flow of genetic information from the nucleus to the translation machinery in the cytoplasm, the nuclear pore complex needs to operate with great precision. Therefore, it is no surprise that a great many diseases have been attributed to nuclear pore dysfunction [18–20]. In certain types of cancers, nucleoporin genes fused to other genes have been reported [18]. Similarly, lethal developmental defects have been reported that are a consequence of heritable mutations in nucleoporin genes [19]. Furthermore, due to the long lifespan of nucleoporins, especially in nondividing cells such as neurons, any damage to these proteins is quite detrimental leading to neurological disorders [20]. Nucleocytoplasmic transport defects have also been reported in Huntington's disease and amyotrophic lateral sclerosis [21–24]. There is accumulating evidence of viruses hijacking the nuclear pore transport system to gain access and control of the nucleus and its functions such as transcription and export of viral mRNA from the nucleus while inhibiting host nucleoplasmic transport [25–27]. While the precise molecular mechanism of viral hijack of nuclear pore transport remains to be determined, intact HIV capsids have been reported to dock via their proline-rich regions to the peptidyl-prolyl isomerization region of nucleoporins at the nuclear pore complex of host cells [28–30]. These observations provide some insight into the possible mechanism of viral infection and the pathophysiology of nuclear pore transport.

References

1. Callan, H. G., Randall, J. T., & Tomlin, S. G. (1949). An electron microscopy study of the nuclear membrane. *Nature, 163*, 280.
2. Callan, H. G., & Tomlin, S. G. (1950). Experimental studies on amphibian oocyte nuclei. I. Investigation of the structure of the nuclear membrane by means of the electron microscope. *Proceedings of the Royal Society of London. Series B-Biological Sciences, 137*, 367–378.
3. Watson, M. L. (1959). Further observations on the nuclear envelope of the animal cell. *The Journal of Biophysical and Biochemical Cytology, 6*, 147–156.
4. Gerace, L., Ottaviano, Y., & Kondor-Koch, C. (1982). Identification of a major polypeptide of the nuclear pore complex. *The Journal of Cell Biology, 95*, 826–837.
5. Davis, L. I., & Blobel, G. (1986). Identification and characterization of a nuclear pore complex protein. *Cell, 45*, 699–709.
6. Hinshaw, J. E., Carragher, B. O., & Milligan, R. A. (1992). Architecture and design of the nuclear pore complex. *Cell, 69*, 1133–1141.
7. Akey, C. W., & Radermacher, M. (1993). Architecture of the Xenopus nuclear pore complex revealed by three-dimensional cryo-electron microscopy. *The Journal of Cell Biology, 122*, 1–19.
8. Cronshaw, J. M., Krutchinsky, A. N., Zhang, W., Chait, B. T., & Matunis, M. J. (2002). Proteomic analysis of the mammalian nuclear pore complex. *The Journal of Cell Biology, 158*, 915–927.
9. Stoffler, D., Feja, B., Fahrenkrog, B., Walz, J., Typke, D., & Aebi, U. (2003). Cryo-electron tomography provides novel insights into nuclear pore architecture: Implications for nucleocytoplasmic transport. *Journal of Molecular Biology, 328*, 119–130.
10. Beck, M., Lucic, V., Forster, F., Baumeister, W., & Medalia, O. (2007). Snapshots of nuclear pore complexes in action captured by cryo-electron tomography. *Nature, 449*, 611–615.
11. Kosinski, J., Mosalaganti, S., von Appen, A., Teimer, R., DiGuilio, A. L., et al. (2016). Molecular architecture of the inner ring scaffold of the human nuclear pore complex. *Science, 352*, 363–365.
12. Kim, S. J., Fernandez-Martinez, J., Nudelman, I., Shi, Y., Zhang, W., et al. (2018). Integrative structure and functional anatomy of a nuclear pore complex. *Nature, 555*, 475–482.
13. Forbes, D. J., Kirschner, M. W., & Newport, J. W. (1983). Spontaneous formation of nucleus-like structures around bacteriophage DNA microinjected into Xenopus eggs. *Cell, 34*(1), 13–23.
14. Lohka, M. J., & Masui, Y. (1984). Roles of cytosol and cytoplasmic particles in nuclear envelope assembly and sperm pronuclear formation in cell-free preparations from amphibian eggs. *The Journal of Cell Biology, 98*(4), 1222–1230.
15. Newmeyer, D. D., Finlay, D. R., & Forbes, D. J. (1986). In vitro transport of a fluorescent nuclear protein and exclusion of non-nuclear proteins. *The Journal of Cell Biology, 103*(6 Pt. 1), 2091–2102.
16. Newport, J. (1987). Nuclear reconstitution in vitro: Stages of assembly around protein-free DNA. *Cell, 48*(2), 205–217.
17. Newport, J., & Dunphy, W. (1992). Characterization of the membrane binding and fusion events during nuclear envelope assembly using purified components. *The Journal of Cell Biology, 116*(2), 295–306.
18. Kohler, A., & Hurt, E. (2010). Gene regulation by nucleoporins and links to cancer. *Molecular Cell, 3*, 6–15.
19. Nousiainen, H. O., Kestila, M., Pakkasjarvi, N., Honkala, H., Kuure, S., Tallila, J., Vuopala, K., Ignatius, J., Herva, R., & Peltonen, L. (2008). Mutations in mRNA export mediator GLE1 result in a fetal motoneuron disease. *Nature Genetics, 40*, 155–157.
20. Savas, J. N., Toyama, B. H., Xu, T., Yates, J. R., 3rd., & Hetzer, M. W. (2012). Extremely long-lived nuclear pore proteins in the rat brain. *Science, 335*, 942.

21. Boehmer, T., Jeudy, S., Berke, I. C., & Schwartz, T. U. (2008). Structural and functional studies of Nup107/Nup133 interaction and its implications for the architecture of the nuclear pore complex. *Molecular Cell, 30*, 721–731.
22. Brohawn, S. G., Leksa, N. C., Spear, E. D., Rajashankar, K. R., & Schwartz, T. U. (2008). Structural evidence for common ancestry of the nuclear pore complex and vesicle coats. *Science, 322*, 1369–1373.
23. Debler, E. W., Ma, Y., Seo, H. S., Hsia, K. C., Noriega, T. R., Blobel, G., & Hoelz, A. (2008). A fence-like coat for the nuclear pore membrane. *Molecular Cell, 32*, 815–826.
24. Brohawn, S. G., & Schwartz, T. U. (2009). Molecular architecture of the Nup84–Nup145C–Sec13 edge element in the nuclear pore complex lattice. *Nature Structural & Molecular Biology, 16*, 1173–1177.
25. Cohen, S., Au, S., & Panté, N. (2011). How viruses access the nucleus. *Biochimica et Biophysica Acta (BBA)-Molecular Cell Research, 1813*, 1634–1645.
26. Le Sage, V., & Mouland, A. J. (2013). Viral subversion of the nuclear pore complex. *Viruses, 5*, 2019–2042.
27. Yarbrough, M. L., Mata, M. A., Sakthivel, R., & Fontoura, B. M. (2014). Viral subversion of nucleocytoplasmic trafficking. *Traffic, 15*, 127–140.
28. Lin, D. H., Zimmermann, S., Stuwe, T., Stuwe, E., & Hoelz, A. (2013). Structural and functional analysis of the C-terminal domain of Nup358/RanBP2. *Journal of Molecular Biology, 425*, 1318–1329.
29. Schaller, T., Ocwieja, K. E., Rasaiyaah, J., Price, A. J., Brady, T. L., et al. (2011). HIV-1 capsid-cyclophilin interactions determine nuclear import pathway, integration targeting and replication efficiency. *PLoS Pathogens, 7*, e1002439.
30. Bichel, K., Price, A. J., Schaller, T., Towers, G. J., Freund, S. M., & James, L. C. (2013). HIV-1 capsid undergoes coupled binding and isomerization by the nuclear pore protein NUP358. *Retrovirology, 10*, 81.

Chapter 7
Myosin: Cellular Molecular Motor

Abstract Myosins are a superfamily of ATP-dependent, actin-based motor proteins. In addition to muscle contraction, myosins are involved in a wide range of motility functions in cells, including transport of intracellular cargo. Typically, myosin molecules are composed of a heavy chain having a tail, hinge, and head domain and light chain present near the myosin head that modulates calcium-dependent transduction of force by myosin. The myosin head has both actin binding and ATP binding and ATPase activity. In resting state, myosin is bound to actin. When ATP binds to the myosin head, the head dissociates from the actin filament. Hydrolysis of bound ATP to ADP and the release of phosphate establish the rigor state of myosin (reestablishes myosin–actin interaction), generating force in the process to walk along the actin filament. Binding of a new ATP molecule to the myosin head releases myosin from actin to repeat the cycle. The structure and function of myosin are globally conserved.

Keywords Myosin · ATP-fueled motor proteins · Actin-based motility

B. P. Jena, *Cellular Nanomachines*,
https://doi.org/10.1007/978-3-030-44496-9_7

Discovery

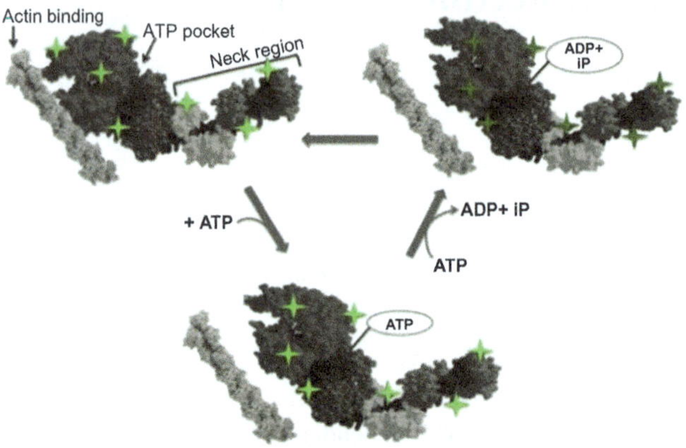

Schematic illustration of myosin-actin interaction utilizing energy in form of ATP. [From *ACS Nano Lett. 2017,* 17: 1262-1268]. ©Bhanu Jena.

In 1859, the molecular motor protein *myosin*, the first cellular nanomachine, was discovered [1]. Only 80 years later in 1939 came the next major breakthrough, when myosin was demonstrated to be an ATPase [2]. In 1941, the first visualization of the contration of myosin filaments in the presence of ATP was made [3]. These studies further suggested the involvement of an additional protein in the ATP-mediated contration of myosin [4]. In 1942–1943, this additional protein was isolated and named *actin*, and muscle contraction was demonstrated to represent an *ATP-fueled actomyosin process* [5, 6]. Subsequently it was discovered that in muscle, actin and myosin are differently localized as thin and thick filaments, respectively, and that movement occurs as a consequence of sliding action between the two filaments [7, 8]. In a 1963 study using electron microscopy (EM) [9] the bipolar structure of the myosin thick filament was demonstrated, and actin thin filaments decorated with myosin heads were observed for the first time. Subsequently, EM studies [10] demonstrated that myosin molecules end in two globules, and that major portion of myosin is composed of two parallel peptide chains. Further EM studies on enzymatically cleaved myosin subfragments revealed the presence of light chains and additionally showed that myosin heads retain full ATPase activity [11]. In a span of 110 years (1859–1969), the skeletal muscle myosin was the first molecular motor to be discovered and its general structure–function revealed at the nanoscale. Today we know that myosins are the most abundant of the three classes of motor proteins, namely myosin, kinesin, and dynein. Eighteen classes of myosins have since been discovered, of which myosin II is the most abundant and found in skeletal muscles.

Structure–Function

Myosin typically has a globular head region (heavy chain) measuring 16.5 nm in length and 6.5 nm in width and a tail (superhelical rod) measuring up to 150 nm in length and 2 nm wide. The myosin head has ATP binding and ATPase activity that utilizes ATP as fuel to perform mechanical work, leading to a wide range of motions, from transport of subcellular organelles within cells to cellular motion, and to the locomotion of an entire organism via skeletal muscle myosin II (Fig. 7.1). Myosin II is a 520 kDa hexamer comprised of two heavy chains and four light chains. Each heavy chain is 220 kDa, composed of a short globular head and a long coiled-coil tail region. The two myosin II light chains modulating force transduction have molecular mass of 22 and 19 kDa, respectively, and belong to a large family of calcium-binding proteins.

The first report that actomyosin ATPase activity can be regulated by direct binding of calcium to myosin was made in 1970 [12]. This led to studies demonstrating that calcium sensitivity of actomyosin ATPase activity could be reconstituted in vitro in the presence of tropomyosin and troponin subunits [13]. With these structural and biochemical understanding of myosin and actin, the "stroke model" [14] and a "four-state model" of force generation in muscle contraction were proposed [15]. Muscle contraction was found to involve ATP hydrolysis cycle, the association and dissociation of actomyosin, and rotation of the cross-bridge [15]. It was additionally found that in the nucleotide-free "rigor" state, myosin head cooperatively activates actin complexes in the absence of calcium [16]. X-ray diffraction studies further demonstrated that upon activation, the myosin heads move toward actin in a sliding manner [17]. Besides calcium, the role of magnesium-dependent ATPase of myosin was also demonstrated, revealing the existence of multiple ATP/ADP intermediate states of myosin [18]. Finally, determination that actomyosin cross-bridge

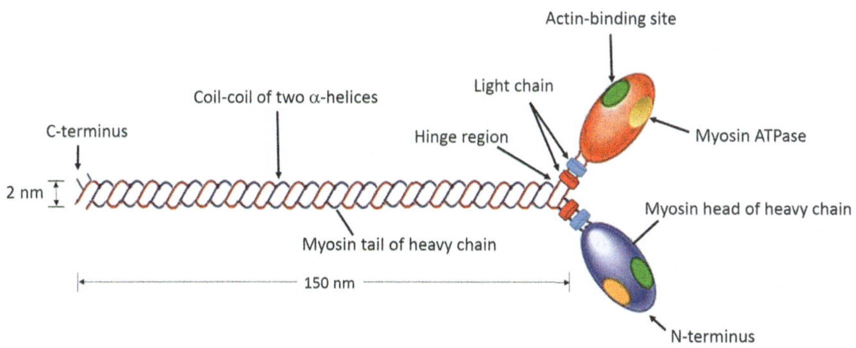

Fig. 7.1 Schematic diagram of skeletal muscle myosin II molecule. Myosin II is composed of two heavy chains, each made up of 2000 amino acids having a tail, hinge, and head domain. The tail length and thickness are 150 nm and 2 nm, respectively. The two light chains of myosin are present near the myosin head and help modulate calcium-dependent transduction of force by the myosin head domain

state is coupled to inorganic phosphate release [19], and the visual demonstration of the movement of fluorescent actin filament on immobilized myosin [20] and S2 region of the myosin head [21], provided the molecular mechanism of force generation by myosin.

New insights into our understanding of myosin function, i.e., determination of the efficiency of myosin as a molecular motor and the role of elevated calcium and magnesium in the process, have come from recent studies [22, 23] utilizing nanothermometry. Greater heat loss when a specific amount of fuel (ATP) is used by a molecular motor reflects decrease in work performed by the motor, hence decreased efficiency. Using zero-dimensional semiconducting cadmium telluride quantum dots (CdTe QDs) that fluoresce green when excited at a wavelength 350 nm (λ_{ex} = 350 nm), losing fluorescence with increasing temperatures, mK-scale nanothermometry have been performed to determine the efficiency of bovine cardiac and rabbit skeletal muscle myosin [22]. Higher the temperature of myosin when hydrolyzing ATP, less the CdTe QD fluorescence. These studies demonstrate that rabbit skeletal muscle myosin is a more efficient molecular motor compared to the bovine cardiac myosin [22].

Similarly, the CdTe-QD-based nanothermometry has been used to gain new insights on the role of elevated calcium and magnesium on myosin structure–function. A 10,000-fold (15–20 mM) higher calcium in the sarcoplasmic reticulum (SR) over what is minimally required for muscle function suggested a physiological role for mM level calcium on muscle function. High calcium levels near active calcium channels at the SR membrane and the impact of localized high calcium on the structure–function of myosin were therefore determined [23]. *Combined CdTe QD-based nanothermometry and circular dichroism (CD) spectroscopy enabled detection of previously unknown enthalpy changes and associated structural remodeling of myosin*, impacting its function following exposure to elevated mM-level calcium (Fig. 7.2) [23]. Exposure of myosin to calcium was demonstrated to be exothermic, resulting in lowering of enthalpy, a decrease in alpha-helical content measured using CD spectroscopy, and the consequent increase in its motor efficiency [23]. Although calcium is the preferred myosin ligand, magnesium too is reported to competitively bind to the motor protein, possibly to the same sites binding calcium. Magnesium is known to modulate actin binding and ADP release in myosin. The CdTe QD-based nanothermometry measurements of myosin in the presence and absence of calcium and magnesium demonstrated that although calcium binds elicit greater thermal shift lowering the energy state of myosin, both divalent cations bind to the same site on the molecule [23]. These results establish a new paradigm in the structure–function of the myosin molecule.

Altered Motor Protein and Disease

A wide range of diseases resulting from alterations in myosin either due to mutation or changes in myosin content and chemistry are known in humans. For example, a mutation in the myosin head domain called K146N causes thickening of the heart

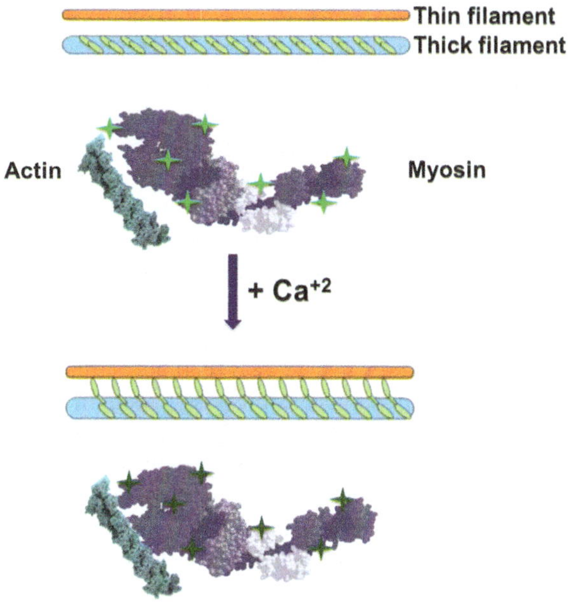

Fig. 7.2 Schematic diagram showing myosin heads in the thick filament unwind and loses enthalpy in the presence of calcium, resulting in increased interactions with the thin filament. (TOP) Thick and thin filaments before calcium addition. Note the bright green fluorescing QDs (bright green stars). (BOTTOM) Calcium addition results in the loss of alpha-helical content of myosin molecules demonstrated using CD and the consequent release of heat and therefore loss of QD fluorescence (dim green stars). Loss in alpha-helical content of the myosin molecule in the presence of calcium could result in a radial increase in size of the molecule, enabling myosin heads greater probability to interaction with actin in the thin filament [23]. ©Bhanu Jena

wall in humans and the same mutation in the fruit fly *Drosophila* exhibits degenerative heart tube and flight muscle phenotype [24]. Similarly, studies in humans and animal model of muscle wasting [25–43] report that muscle immobilization results primarily in the loss of skeletal muscle myosin II and its altered post-translational modifications. Skeletal muscles in adult mammals predominantly express myosin's MyHC-I, -IIa, -IIx/d, and -IIb. Although adult human skeletal muscle does not express the MyHC-IIb isoform, its expression pattern has been demonstrated to be influenced by exercise, disease, and age [44–47]. Quantification of MyHC mRNA demonstrates that transcripts for several isoforms could be present in a single fiber and the expression of myosin isoforms would vary between fibers [48–51]. Therefore, changes in fiber composition of myosin isoforms could occur in case of immobilization-induced skeletal muscle myopathy. Similarly, since myosin molecules have a slow turnover rate of 1–2% per day [52], they are targets for post-translational modifications (PTMs) by reactive oxygen or nitrogen species [25], serving as indicators for assessing immobilization-induced skeletal muscle myopathy. PTMs have therefore been reported and provide dramatic alterations to both the myosin protein structure and its function [52–61]. In studies [19] on human skeletal muscle tissue obtained from young (20–35 years) and older subjects (71–91 years), mass spectrometry analyses combined with X-ray diffraction and motility measurements find eight age-specific myosin PTMs, in which two were located in the motor domain (carbonylation of Pro79 and Asn81) and six in the rod region (carbonylation of Asp900, Asp904, and Arg908; methylation of Glu1166; deamidation of Gln1164 and Asn1168) of myosin. This study [19] further revealed that PTMs in the motor domain were only present in the IIx/d MyHC isoform, suggesting PTMs in the rod region contributed to the observed disordering of myosin filaments and the slowing of motility speed and reduced force generation capacity. Similarly, gene profiling studies on skeletal muscle and myosin protein expression, PTMs of myosin and single muscle fiber motility assays, demonstrate a preferential loss of myosin and myosin associated proteins, impaired force generation capacity, and changes in myosin in ICU patients. Temporal gene expression pattern in human muscle biopsies obtained from healthy and ICU patients [27], and from the experimental rat ICU model, suggests that complete mechanical silencing (lack of weight-bearing) is an important trigger of critical illness myopathy in ICU patients [30]. A striking early and maintained upregulation (6 hours to 14 days) of muscle atrogenes (muscle ring-finger 1/tripartite motif-containing 63 and F-box protein 32/atrogin-1) is observed, followed by an upregulation of other protein degradation pathways at intermediate and long durations (5 days–14 days) of mechanical silencing. Oxidative stress response genes and genes that participate in amino acid catabolism, cell cycle arrest, apoptosis are found to be significantly upregulated from 5 to 14 days following mechanical silencing, and genes related to contractile (myosin heavy chain and myosin binding protein C), regulatory (troponin, tropomyosin), and mitochondrial proteins were downregulated [30]. The myosin loss and muscle wasting follow a temporal pattern, with an initial sparing of both muscle function, mass and myosin content, accompanied by a progressive loss of muscle force that exceeds the loss in muscle mass due to a preferential loss of the motor protein myosin [27, 30, 62].

Alongside non-invasive clinical examination, bioelectrical impedance analysis, magnetic resonance spectroscopy, ultrasonography, computed tomography, and X-ray absorptiometry [63–69] are among the approaches used to diagnose myopathy. However, skeletal muscle biopsy commonly used for light microscopic, electron microscopic, and biochemical and transcriptional evaluation of the skeletal muscle remains the gold standard for establishing the etiology of a myopathy [70–74]. Myopathy observed in light and electron micrographs of biopsy tissue includes focal myofiber damage as in mitochondrial disorders, segmental damage in dystrophies, or multifocal damage in various inflammatory myopathies [75, 76]. Common myopathic features include variable fiber size with both atrophied and hypertrophied muscle fibers. Atrophied fibers are often rounded, as opposed to the angulated atrophic fibers observed in neurogenic myopathy. Certain myopathies exhibit ragged red fibers which are common in mitochondrial disease. Other myopathies exhibit central core structures found in central core disease and the presence of rimmed vacuoles found in inclusion body myositis [77–79]. While most myopathies exhibit one or more of such phenotypes, the very early stages or several metabolic myopathies often exhibit normal muscle morphology on routine histochemical examination, making diagnosis difficult. In such cases where standard staining techniques fail to offer definitive diagnostic information, a combination of expensive and time-consuming immunohistochemistry, electron microscopy, and biochemical testing is required to provide definitive diagnosis of the disease and for its appropriate treatment and management. To overcome these limitations in the diagnostic pathology of skeletal muscle tissue, we have developed an accurate, inexpensive, and nanoscale imaging approach combined with computational image analysis and machine learning which we refer to as differential expansion microscopy (DiExM) [80, 81]. Therefore, the optimized application of DiExM 3D imaging using stochastic optical reconstruction microscopy, combined with biochemical and functional markers, to establish an artificial intelligence-driven platform will help in the determination of various stages of immobilization-induced skeletal muscle myopathies in humans. The development of a collection of neural network architectures capable of deciphering with great precision the various stages of immobilization-induced skeletal muscle myopathy will form the cornerstone in establishing the next generation of diagnosis of human myopathies.

References

1. Kühne, W. (1859). Untersuchungen über bewegungen und veränderungen der kontraktileu substanzen. *Archiv für Anatomie. Physiologie und wissenschaftliche Medicin, 1859*, 748–783.
2. Engelhardt, W. A., & Ljubimowa, M. N. (1939). Myosine and adenosine triphosphate. *Nature, 144*, 668–669.
3. Banga, I., & Szent-Gyorgyi, A. (1941). Preparation and properties of myosin A and B. In *Studies from the Institute of Medical Chemistry University of Szeged, edited by Szent-Gyorgyi A* (Vol. I, pp. 5–15). New York: Karger.
4. Banga, I., & Szent-Gyorgyi, A. (1943). The influence of salts on the phosphatase action of myosin. In A. Szent-Gyorgyi (Ed.), *Studies from the Institute of Medical Chemistry University of Szeged* (Vol. III, pp. 72–75). New York: Karger.
5. Straub, F. B. (1942). Actin. In A. Szent-Gyorgyi (Ed.), *Studies from the Institute of Medical Chemistry University of Szeged* (Vol. II, pp. 3–15). New York: Karger.
6. Straub, F. B. (1943). Actin II. In *Studies from the Institute of Medical Chemistry University of Szeged, edited by Szent-Gyorgyi A* (Vol. III, pp. 23–37). New York: Karger.
7. Huxley, A. F., & Niedergerke, R. (1954). Structural changes in muscle during contraction: Interference microscopy of living muscle fibres. *Nature, 173*, 971–973.
8. Huxley, H. E., & Hanson, J. (1954). Changes in the cross-striations of muscle during contraction and stretch and their structural interpretation. *Nature, 173*, 973–976.
9. Huxley, H. E. (1963). Electron microscope studies on the structure of natural and synthetic protein filaments from striated muscle. *Journal of Molecular Biology, 7*, 281–308.
10. Slayter, H. S., & Lowey, S. (1967). Substructure of the myosin molecule as visualized by electron micros- copy. *Proceedings of the National Academy of Sciences, 58*, 1611–1618.
11. Lowey, S., Slayter, H. S., Weeds, A. G., & Baker, H. (1969). Substructure of the myosin molecule. I. Subfragments of myosin by enzymic degradation. *Journal of Molecular Biology, 42*, 1–29.
12. Kendrick-Jones, J., Lehman, W., & Szent-Gyorgyi, A. G. (1970). Regulation in molluscan muscles. *Journal of Molecular Biology, 54*, 313–326.

13. Greaser, M. L., & Gergely, J. (1971). Reconstitution of troponin activity from three protein components. *The Journal of Biological Chemistry, 246*, 4226–4233.
14. Huxley, A. F., & Simmons, R. M. (1971). Proposed mechanism of force generation in striated muscle. *Nature, 233*, 533–538.
15. Lymn, R. W., & Taylor, E. W. (1971). Mechanism of adenosine triphosphate hydrolysis by actomyosin. *Biochemistry, 10*, 4617–4624.
16. Bremel, R. D., & Weber, A. (1972). Cooperation within actin filament in vertebrate skeletal muscle. *Nature: New Biology, 238*, 97–101.
17. Haselgrove, J. C., & Huxley, H. E. (1973). X-ray evidence for radial cross-bridge movement and for the sliding filament model in actively contracting skeletal muscle. *Journal of Molecular Biology, 77*, 549–568.
18. Bagshaw, C. R., & Trentham, D. R. (1974). The characterization of myosin-product complexes and of product-release steps during the magnesium ion-dependent adenosine triphosphatase reaction. *The Biochemical Journal, 141*, 331–349.
19. Hibberd, M. G., Dantzig, J. A., Trentham, D. R., & Goldman, Y. E. (1985). Phosphate release and force generation in skeletal muscle fibers. *Science, 228*, 1317–1319.
20. Kron, S. J., & Spudich, J. A. (1986). Fluorescent actin filaments move on myosin fixed to a glass surface. *Proceedings of the National Academy of Sciences, 83*, 6272–6276.
21. Toyoshima, Y. Y., Kron, S. J., McNally, E. M., Niebling, K. R., Toyoshima, C., & Spudich, J. A. (1987). Myosin subfragment-1 is sufficient to move actin filaments in vitro. *Nature, 328*, 536–539.
22. Laha, S., Naik, A. R., Kuhn, E. R., Alvarez, M., Sujkowski, A., Wessels, R. J., & Jena, B. P. (2017). Nanothermometry measure of muscle efficiency. *ACS Nano Letters, 17*, 1262–1268.
23. Kuhn, E. R., Naik, A. R., Lewis, B. E., Kokotovich, K. M., Li, M., Stemmler, T. M., et al. (2018). Nanothermometry reveals calcium-induced remodeling of myosin. *ACS Nano Letters, 18*, 7021–7029.
24. Kronert, W. A., Bell, K. M., Viswanathan, M. C., Melkani, G. C., Trujillo, A. S., Huang, A., et al. (2018). Prolonged cross-bridge binding triggers muscle dysfunction in a fly model of myosin-based hypertrophic cardiomyopathy. *eLife, 7*, e38e064.
25. Li, M., Ogilvie, H., Ochala, J., Konstantin, A., Iwamoto, H., Yagi, N., et al. (2015). Aberrant post-translational modifications compromise human myosin motor function in old age. *Aging Cell, 14*, 228–235.
26. Llsano-Diez, M., Renaud, G., Andersson, M., Marrero, H. G., Cacciani, N., Engquist, H., Corpeno, R., Artemenko, K., Bergquist, J., & Larsson, L. (2012). Mechanism underlying ICU muscle wasting and effects of passive mechanical loading. *Critical Care, 16*, R209.
27. Llsano-Diez, M., Gustafsin, A.-M., Olsson, C., Goransson, H., & Larsson, L. (2011). Muscle wasting and the temporal gene expression pattern in a novel rat intensive care unit model. *BMC Genomics, 12*, 602.
28. Norman, H., Nordquist, J., Andersson, P., Ansved, T., Tang, X., Dworkin, B., et al. (2006). Impact of post-synaptic block of neuromuscular transmission, muscle unloading and mechanical ventilation on skeletal muscle protein and mRNA expression. *Pflügers Archiv, 453*, 53–66.
29. Larsson, L. (2007). Experimental animal models of muscle wasting in intensive care unit patients. *Critical Care Medicine, 35*(9 Suppl), S484–S487.
30. Ochala, J., Gustafson, A.-M., Li, M., Aare, S., Qaisar, R., Llano Diez, M., et al. (2011). Preferential skeletal muscle myosin loss in response to mechanical silencing in a novel rat intensive care unit model: underlying mechanisms. *The Journal of Physiology, 589*(8), 2007–2026.
31. Akkad, H., Corpeno, R., & Larsson, L. (2014). Masseter muscle myofibrillar protein synthesis and degradation in an experimental critical illness myopathy model. *Plos-One., 9*(4), e92622.
32. Pette, D., & Staron, R. S. (2000). Myosin isoforms, muscle fiber types, and transitions. *Microscopy Research and Technique, 50*(6), 500–509.
33. MacFarlane, I. A., & Rosenthal, F. D. (1977). Severe myopathy after status asthmaticus. *Lancet, 2*(8038), 615.

34. Lacomis, D., Zochodne, D. W., & Bird, S. J. (2000). Critical illness myopathy. *Muscle & Nerve, 23*(12), 1785–1788.
35. Phillips, S. M., Glover, E. I., & Rennie, M. J. (2009). Alterations of protein turnover underlying disuse atrophy in human skeletal muscle. *Journal of Applied Physiology, 107*(3), 645–654.
36. Cohen, S., Brault, J. J., Gygi, S. P., et al. (2009). During muscle atrophy, thick, but not thin, filament components are degraded by MuRF1-dependent ubiquitylation. *The Journal of Cell Biology, 185*(6), 1083–1095.
37. Lecker, S. H., Jagoe, R. T., Gilbert, A., et al. (2004). Multiple types of skeletal muscle atrophy involve a common program of changes in gene expression. *The FASEB Journal, 18*(1), 39–51.
38. Sacheck, J. M., Hyatt, J. P., Raffaello, A., et al. (2007). Rapid disuse and denervation atrophy involve transcriptional changes similar to those of muscle wasting during systemic diseases. *The FASEB Journal, 21*(1), 140–155.
39. Sandri, M. (2010). Autophagy in health and disease. 3. Involvement of autophagy in muscle atrophy. *American Journal of Physiology. Cell Physiology, 298*(6), C1291–C1297.
40. Stevenson, E., Giresi, P., Koncarevic, A., & Kandarian, S. (2003). Global analysis of gene expression patterns during disuse atrophy in rat skeletal muscle. *Journal of Physiology-London., 2003*, 33–48.
41. Sasa, T., Sairyo, K., Yoshida, N., et al. (2004). Continuous muscle stretch prevents disuse muscle atrophy and deterioration of its oxidative capacity in rat tail-suspension models. *American Journal of Physical Medicine & Rehabilitation, 83*(11), 851–856.
42. Sandri, M., Lin, J., Handschin, C., Yang, W., Arany, Z. P., Lecker, S. H., et al. (2006). PGC-1alpha protects skeletal muscle from atrophy by suppressing FoxO3 action and atrophy- specific gene transcription. *Proceedings of the National Academy of Sciences of the United States of America, 103*, 16260–16265.
43. Ruas, J. L., White, J. P., Rao, R. R., Kleiner, S., Brannan, K. T., Harrison, B. C., et al. (2012). A PGC-1alpha isoform induced by resistance training regulates skeletal muscle hypertrophy. *Cell, 151*, 1319–1331.
44. Resnicow, R., Deacon, J. C., Warrick, H. M., Spudich, J. A., & Leinwand, L. A. (2010). Functional diversity among a family of human skeletal muscle myosin motors. *Proceedings of the National Academy of Sciences of the United States of America, 107*, 1053–1058.
45. Schiaffino, S., & Reggiani, C. (1994). Myosin isoforms in mammalian skeletal muscle. *Journal of Applied Physiology, 77*, 493–501.
46. Schiaffino, S., & Reggiani, C. (1996). Molecular diversity of myofibrillar proteins: Gene regulation and functional significance. *Physiological Reviews, 76*, 371–423.
47. Pette, D., & Staron, R. S. (1997). Mammalian skeletal muscle fiber type transitions. *International Review of Cytology, 170*, 143–223.
48. Staron, R. S., & Pette, D. (1987). Nonuniform myosin expression along single fibers of chronically stimulated and contralateral rabbit tibialis anterior muscles. *Pflügers Archiv, 409*, 67–73.
49. Peuker, H., & Pette, D. (1997). Quantitative analyses of myosin heavy-chain mRNA and protein isoforms in single fibers reveal a pronounced fiber heterogeneity in normal rabbit muscles. *European Journal of Biochemistry, 247*, 30–36.
50. Andersen, J. L., & Schiaffino, S. (1997). Mismatch between myosin heavy chain mRNA and protein distribution in human skeletal muscle fibers. *American Journal of Physiology-Cell Physiology, 272*, C1881–1889. 37.
51. Eizema, K., et al. (2003). Differential expression of equine myosin heavy-chain mRNA and protein isoforms in a limb muscle. *The Journal of Histochemistry and Cytochemistry, 51*, 1207–1216.
52. Smith, K., & Rennie, M. J. (1996). The measurement of tissue protein turnover. *Baillière's Clinical Endocrinology and Metabolism, 10*, 469–495.
53. Mann, M., & Jensen, O. (2003). Proteomic analysis of post-translational modifications. *Nature Biotechnology, 21*, 255–261.
54. Kim, W., Bennett, E. J., Huttlin, E. L., Guo, A., Li, J., Possemato, A., et al. (2011). Systematic and quantitative assessment of the ubiquitin-modified proteome. *Molecular Cell, 44*, 325–340.

55. Edwards, A. V. G., Cordwell, S. J., & White, M. Y. (2011). Phosphoproteomic pro ling of the myocyte. *Circulation. Cardiovascular Genetics, 4*, 575.
56. Hennet, T. (2012). Diseases of glycosylation beyond classical congenital disorders of glycosylation. *Biochimica et Biophysica Acta, 1820*, 1306–1317.
57. Li, T., Evdokimov, E., Shen, R.-F., Chao, C.-C., Tekle, E., Wang, T., Stadtman, E. R., Yang, D. C. H., & Chock, P. B. (2004). Sumoylation of heterogeneous nuclear ribonucleoproteins, zinc finger proteins, and nuclear pore complex proteins: A proteomic analysis. *Proceedings of the National Academy of Sciences of the United States of America, 101*, 8551–8556.
58. Matafora, V., D'Amato, A., Mori, S., Blasi, F., & Bachi, A. (2009). Proteomics analysis of nucleolar SUMO-1 target proteins upon proteasome inhibition. *Molecular & Cellular Proteomics, 8*, 2243–2255.
59. Flick, K., & Kaiser, P. (2009). Proteomic revelation: SUMO changes partners when the heat is on. *Science Signaling, 2*, pe45.
60. Tatham, M. H., Rodriguez, M. S., Xirodimas, D. P., & Hay, R. T. (2009). Detection of protein SUMOylation in vivo. *Nature Protocols, 4*, 1363–1371.
61. Galisson, F., Mahrouche, L., Courcelles, M., Bonneil, E., Meloche, S., Chelbi-Alix, M. K., & Thibault, P. (2011). A novel proteomics approach to identify SUMOylated proteins and their modification sites in human cells. *Molecular & Cellular Proteomics, 10*, M110.004796.
62. Norman, H., Nordquist, J., Andersson, P., Ansved, T., Tang, X., Dworkin, B., & Larsson, L. (2006). Impact of post-synaptic block of neuromuscular transmission, muscle unloading and mechanical ventilation on skeletal muscle protein and mRNA expression. *Pflügers Archiv, 453*, 53–66.
63. Lee, S. Y., Ahn, S., Kim, Y. J., Ji, M. J., Kim, K. M., Choi, S. H., et al. (2018). Comparison between dual-energy X-ray absorptiometry and bioelectrical impedance analyses for accuracy in measuring whole body muscle mass and appendicular skeletal muscle mass. *Nutrients, 10*, 738.
64. Mitsiopoulos, N., Baumgartner, R. N., Heymsfield, S. B., Lyons, W., Gallagher, D., & Ross, R. (1998). Cadaver validation of skeletal muscle measurement by magnetic resonance imaging and computerized tomography. *Journal of Applied Physiology, 85*, 115–122.
65. Reeves, N. D., Maganaris, C. N., & Narici, M. V. (2004). Ultrasonographic assessment of human skeletal muscle size. *European Journal of Applied Physiology, 91*, 116–118.
66. Mourtzakis, M., Parry, S., Connolly, B., & Puthucheary, Z. (2017). Skeletal muscle ultrasound in critical care: A tool in need of translation. *Annals of the American Thoracic Society, 14*, 1495–1503.
67. Mercuri, E., Pichiecchio, A., Allsop, J., Messina, S., Pane, M., & Muntoni, F. (2007). Muscle MRI in inherited neuromuscular disorders: Past, present, and future. *Journal of Magnetic Resonance Imaging, 25*(2), 433–440.
68. Mercuri, E., Jungbluth, H., & Muntoni, F. (2005). Muscle imaging in clinical practice: Diagnostic value of muscle magnetic resonance imaging in inherited neuromuscular disorders. *Current Opinion in Neurology, 18*(5), 526–537.
69. Zaidman, C. M., Holland, M. R., Anderson, C. C., & Pestronk, A. (2008). Calibrated quantitative ultrasound imaging of skeletal muscle using backscatter analysis. *Muscle & Nerve, 38*(1), 893–898.
70. Lai, C. H., Melli, G., Chang, Y. J., Skolasky, R. L., Corse, A. M., Wagner, K. R., & Cornblath, D. R. (2010). Open muscle biopsy in suspected myopathy: Diagnostic yield and clinical utility. *European Journal of Neurology, 17*(1), 136–142.
71. Collins, M. P., Mendell, J. R., Periquet, M. I., Sahenk, Z., Amato, A. A., Gronseth, G. S., et al. (2000). Superficial peroneal nerve/peroneus brevis muscle biopsy in vasculitic neuropathy. *Neurology, 55*(5), 636–643.
72. Tarnopolsky, M. A., Pearce, E., Smith, K., & Lach, B. (2011). Suction-modified Bergström muscle biopsy technique: Experience with 13,500 procedures. *Muscle & Nerve, 43*(5), 717–725.

73. Giagnacovo, M., Cardani, R., Meola, G., Pellicciari, C., & Malatesta, M. (2010). Routinely frozen biopsies of human skeletal muscle are suitable for morphological and immunocytochemical analyses at transmission electron microscopy. *European Journal of Histochemistry, 54*(3), e31.
74. Bossen, E. (2000). In R. Wortmann (Ed.), *Muscle biopsy in disease of skeletal muscle* (pp. 333–348). Philadelphia, PA: Lippincott Williams and Wilkins.
75. Dalakas, M. C. (2002). Muscle biopsy findings in inflammatory myopathies. *Rheumatic Diseases Clinics of North America, 28*(4), 779–798.
76. Scola, R. H., Pereira, E. R., Lorenzoni, P. J., & Werneck, L. C. (2007). Toxic myopathies: Muscle biopsy features. *Arquivos de Neuro-Psiquiatria, 65*(1), 82–86.
77. Rifai, Z., Welle, S., Kamp, C., & Thornton, C. A. (1995). Ragged red fibers in normal aging and inflammatory myopathy. *Annals of Neurology, 37*(1), 24–29.
78. Jungbluth, H., Sewry, C. A., & Muntoni, F. (2011). Core myopathies. *Seminars in Pediatric Neurology, 18*(4), 239–249.
79. Huizing, M., & Krasnewich, D. M. (2009). Hereditary inclusion body myopathy: A decade of progress. *Biochimica et Biophysica Acta, 1792*(9), 881–887.
80. Pernal, S.P., Liyanaarachchi, A., Gatti, D.L., Formosa, B., Pulvender, R., Kuhn, E.R., et al. (2019) Differential expansion microscopy. *bioRxiv*. doi:https://doi.org/10.1101/699579.
81. Gatti, D.L., Arslanturk, S., Lal, S., Jena, B.P. (2019) Deep learning strategies for differential expansion microscopy. *bioRxiv*. doi:https://doi.org/10.1101/743682.

Chapter 8
Assembly of Cellular Nanomachines

Chirality and Enthalpy Drive Assembly of Higher Order Cellular Structures

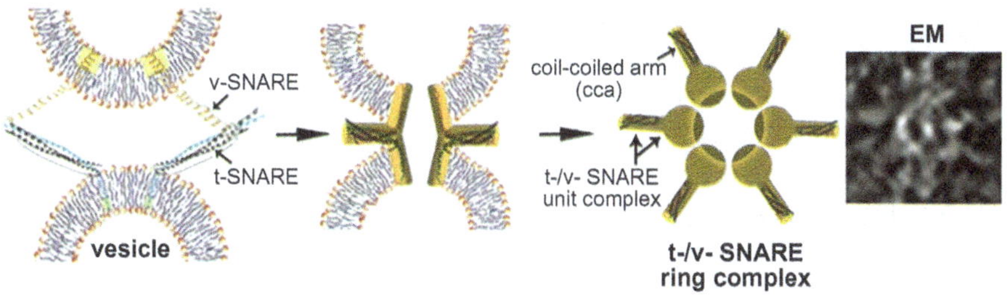

Figure: *Schematic diagram and the EM micrograph of the self-assembled t-/v--SNARE ring complex established when v-SNARE-associated membrane interacts with t-SNARE-associated membrane [Taken from J. Cell. Mol. Med. 2011, 15:31–37.]* ©*Bhanu Jena.*

The unifying principle of all cellular nanomachines is their precisely assembled nanoscale structures and their utilization of energy to operate. So how are cellular nanomachines assembled? What triggers the initiation and termination of their assembly? These are some of the least understood fundamental open questions. In this chapter, an attempt to explore some of the operating principles that may govern the assembly of cellular higher order structures has been made.

Cells are composed of tens of millions of proteins, lipids, nucleic acids, and ions, all solvated and partitioned using billions of water molecules, the most abundant molecule in living organisms. Electron spin alignment and the associated exchange energy in chiral molecules are known to impact biomolecular structure–function and assembly. Although a single water molecule exhibits symmetry and zero chirality, structures formed by water molecules utilizing hydrogen bonding can possess chirality. Though the atomic structure of a wide range of biomolecular complexes have been solved, their initiation and termination remain a mystery. However, in the past two decades, nature-inspired strategies that have been utilized to build synthetic higher ordered structures and nanomachines provide new insights into the assembly of supramolecular complexes in cells. Among such directed self-assembled

B. P. Jena, *Cellular Nanomachines*,
https://doi.org/10.1007/978-3-030-44496-9_8

cellular structures is the membrane-directed self-assembly of the t-/v-SNARE ring complexes critical for membrane fusion and the incorporation of new membrane to pre-existing cellular membranes. Results from these studies provide clues that the major governing principle in the assembly of cellular supramolecular structures is chirality and enthalpy. The energy landscape and domain-specific interactions through hydrogen bonding, electrostatic interactions, and van der Waals forces between biomolecules play critical roles leading to changes in phase properties of biomolecules and molecular complexes that drive the directed assembly of higher order structures in cells.

All matter obey the Pauli exclusion principle. Accordingly, in an atom, a maximum of two electrons can exist in an orbital only if each electron has opposite spins. This split energy is small; however, the stabilization energy is much greater when two electrons of opposite spins occupy the same volume. Hence, spin alignment can control chemical reactions. Consequently, chiral-induced electron spin selectivity effects show that spin orientation is associated with the symmetry of molecules. Therefore, spin alignment, and the associated exchange energy in chiral molecules, impacts biomolecular structure–function and assembly [1, 2]. Although a single water molecule exhibits symmetry and zero chirality, a structure formed by water molecules utilizing hydrogen bonding can possess chirality. DNA's chiral spine of hydration is a good example of the chiral nature of a water superstructure [3]. At the molecular level in cells, proteins are composed exclusively of L-amino acids, while nucleic acids have only D-sugars. Without this chiral asymmetry, biologically active polymers and supramolecular complexes would most likely not have evolved [4, 5]. A major unsolved problem is worth noting as to why is it that we see no D-amino acids in tissues of living organisms (homochirality) when a racemic mixture of equal amounts of both L- and D-amino acids (with the exception of glycine) is produced in vitro, in the absence of some form of a directing template. While various models have been proposed for the evolutionary basis of homochirality in living organisms [6], "symbiotic evolution" may have and continue to play a major role. It is well known that glycosylated polymer of D-alanine and D-glutamic acid are essential components of the cell wall in bacteria and is one of the most important sources of D-amino acids in nature, released by bacteria [7–11]. This D-amino acid-containing peptidoglycan polymer prevents the hosts L-amino acid proteases to cleave bacterial cell wall. Similarly, D-amino acids have been reported to trigger biofilm disassembly [10] and D-tryptophan from probiotic bacteria influences the gut microbiome and impacts immune response in the airways [11].

Chirality in the peptide-induced formation of cholesterol-rich domains is well known [12]. Taking lessons from cellular rotary motors such as ATP synthase and vH^+-ATPase, synthetic molecular motors have been built and confirmed directionality, utilizing intrinsic asymmetry and/or specific sequence of chemical transformations. The structure and arrangement of molecular motors providing either linear, rotary, oscillatory, or reciprocating motions also provide precision in the regulation of the direction of motion. A recent study [13] has elegantly utilized a chiral guest molecule to confer changeable directionality to a photoswitchable photon-driven synthetic molecular motor. Although chemical-energy driven motors are governed by microscopic reversibility [14–16] and light-driven motors [17, 18] governed by the Bose–Einstein relation for absorption and emisson of photons, the novel utilization of a chiral guest molecule to confer changeable directionality to a photon-driven synthetic molecular motor provides new insights into supramolecular assembly and function of cellular rotary motors. The chiral guest molecule induces unidirectional rotational motion around the double bond in the photoswitchable receptor complex [13]. Since the rotational direction is determined by the chirality of the guest molecule, such guest molecules could be selectively changed to reverse the direction of the rotary motion. Cellular rotary motors such as ATP synthase, the cellular nanomachine for ATP biosynthesis, exhibit clockwise rotation (viewed from the opposite side of the matrix), while vH^{+}-ATPase the proton pump that utilizes ATP to pump protons exhibits reverse counter-clockwise rotary motion. ATP synthase is also capable of counter-clockwise rotary motion while hydrolyzing ATP instead of synthesis. The thermodynamic balance between the protomotive force and the phosphorylation potential of ATP guides rotation direction. The direction of rotation in ATP synthase is dictated by the conserved steric arrangement of the chiral "a" subunit of Fo [19].

All higher ordered supramolecular structures in cells are assembled in steps, resulting in the formation of intermediate structures. An understanding of such intermediate structures is critical to gain insights on cellular supramolecular assemblies. However, the instability of such intermediates has made it extremely difficult to capture their structure and assembly processes. To overcome this limitation, a recent study [20] has utilized a thermal hysteresis (TH) approach based on rapid nonequlibrium melting and annealing measurements, to map the rate of supramolecular assembly as a function of temperature and concentration. In this study [20], analysis of the molecular surfaces during assembly of the polydeoxyadenosine (poly(A)) and cyanuric acid fiber complex provides information on the nature of the assembly pathway and the energy landscape of the supramolecular fiber assembly. Results from the study [20] demonstrate that fiber elongation is initiated when an unstable intermediate complex composed of three poly(A) monomers acquires a fourth strand. Similarly, TH has also been used to determine two-state folding and unfolding properties and rates of biomolecules [18, 21]. Spectroscopic signatures of folding and assembly of biomolecules can be gained while raising and lowering the reaction temperature to induce melting and annealing of the complex [18, 21]. In such TH study of two-state folding process, a temperature scan rate is selected that is faster compared to the time required for the system to reach equilibrium, so that folding and unfolding of the complex occur out of equilibrium. Under such rapidly changing temperatures, a 50% folding point or T_m is reached at a higher temperature than the true required T_m on the up-scan and at a lower temperature than the true T_m on the down-scan. This enables the calculation of folding and unfolding rates as a function of temperature based on the size of the lag [18].

There is growing evidence that the basic principles governing cellular supramolecular assembly are chirality and enthalpy of molecules guiding domain-specific interactions through hydrogen bonding, electrostatic, pi–pi, and van der Waals interactions, and the consequent changes in phase properties of molecules and molecular complexes [21–28] in cells [29–35]. Among the directed self-assembly of higher order cellular structures is the membrane-directed self-assembly of the t-/v-SNARE ring complexes involved in membrane fusion in cells, the addition of new cellular membranes to pre-existing membranes, and the bridging of calcium to the oxygens of phospholipid head groups between opposing bilayers to promote membrane fusion. Studies on these critical cellular processes have been conducted in the authors' laboratory [30–41] and presented here as examples of directed supramolecular assembly in cells.

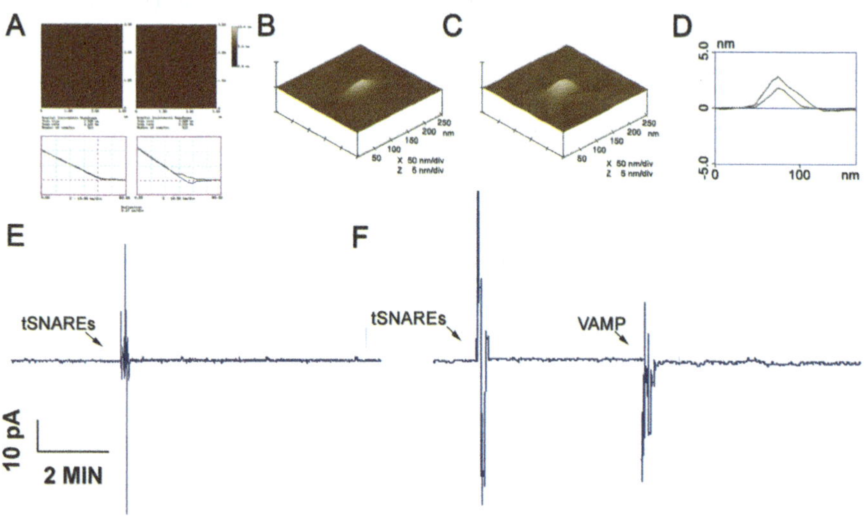

Fig. 8.1 In the absence of membrane, t-SNAREs and v-SNARE fail to form membrane fusion complexes [30]. AFM performed on freshly cleaved mica surface (**a, left**), and on lipid membrane formed on the same mica surface (**a, right**), demonstrating differences in the force vs. distance curves. Note the curvilinear shape exhibited in the force vs. distance plot on the lipid surface in contrast to mica. Three-dimensional AFM micrographs of full-length recombinant neuronal t-SNAREs deposited on the lipid membrane (**b**) and following the addition of a full-length recombinant neuronal v-SNARE (**c**). Section analysis of the SNARE complex in (**b**) and (**c**) is depicted in (**d**). Note the smaller globular structure belonging to the t-SNARE complex in (**b**) is markedly enlarged following the addition of v-SNARE in (**c**). Bilayer lipid membranes are nonconducting either in the presence or absence of SNAREs (**e, f**). Current-time traces of the bilayer membrane in an EPC9 bilayer set up, while the membrane is held at −60 mV (current/reference voltage). (**e**) When t-SNAREs (syntaxin 1A-1 and SNAP25) are added to the planar lipid membrane, no current is observed ($n = 7$). (**f**) Similarly, no current spikes are observed when full-length non-membrane associated v-SNAREs are added to the t-SNARE-reconstituted bilayer (taken from Cho et al., ref. 30). ©Bhanu Jena

Membrane-Directed Self-Assembly of t-/v-SNARE Ring Complexes

Fusion between opposing membrane in cells occurs via soluble *N-ethylmaleimide*-sensitive factor (NSF)-attachment protein receptors called SNAREs [42, 43]. SNAREs are classified as v-SNARE and t-SNAREs, depending on their primary location either in vesicles (v-) or in target (t-) membranes [42, 43]. T- and v-SNAREs reconstituted into lipid vesicle membranes can promote fusion of the opposing membranes, suggesting SNAREs to be the minimal membrane fusion machinery [43]. Imaging studies using atomic force microscopy (AFM) on SNAREs [30, 31] in physiological buffered solutions demonstrate that full-length recombinant t-SNAREs form globular-shaped structure when applied to a lipid bilayer membrane (Fig. 8.1a, b). Addition of full-length recombinant v-SNAREs to this membrane-associated t-SNARE results in little change to the globular-shaped t-SNARE structure (Fig. 8.1c, d). However, if the membrane-associated t-SNARE globular-shaped structure is exposed to membrane-associated v-SNAREs, as in v-SNARE-vesicles, a conducting t-/v-SNARE ring complex is established, meaning the opposing membranes have fused to establish a fusion pore (Fig. 8.2) [30]. Furthermore, the size of the t-/v-SNARE ring complex is directly proportional to the size of the v-SNARE-associated liposome [31]. Therefore, membrane curvature dictates the size of the t-/v-SNARE ring complex (Fig. 1.15, Chap. 1) [30–35]. The built-in enthalpy in the different SNARE molecules dictated by their amino acid sequence and their membrane association is utilized to enable two opposing bilayers to come close together overcoming repulsive forces between the opposing membrane bilayers, promoting membrane fusion in the presence of calcium. Once the t-/v-SNARE ring complex has been established, energy in the form of ATP and the

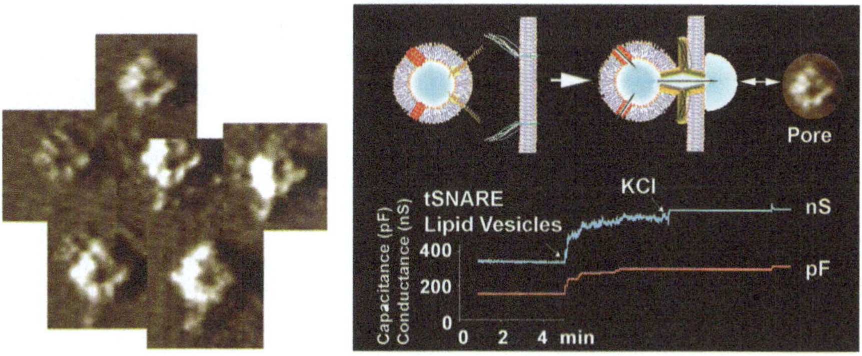

Fig. 8.2 Membrane-associated t- and v-SNAREs interact in a circular array to form conducting pores (fusion pores) [30]. (**Left**) AFM micrograph of t-/v-SNARE ring complexes formed when liposome-associated v-SNAREs interact with membrane-associated t-SNAREs. (**Right**) Note the conductance and capacitance increase following v-SNARE vesicle fusion at the t-SNARE associated membrane (modified figure taken from Cho et al., ref. 30). ©Bhanu Jena

molecular motor protein NSF an ATPase, is required to disassemble the t-/v-SNARE ring complex [32].

Biogenesis of Cellular Membranes

The cell plasma membrane is a highly dynamic organelle governing a wide range of cellular activities including ion transport, secretion, cell division, growth, and development. The fundamental process involved in the addition of new membrane to pre-existing plasma membrane has recently been determined from studies using biophysical, biochemical, and molecular dynamic simulations [35]. Results from

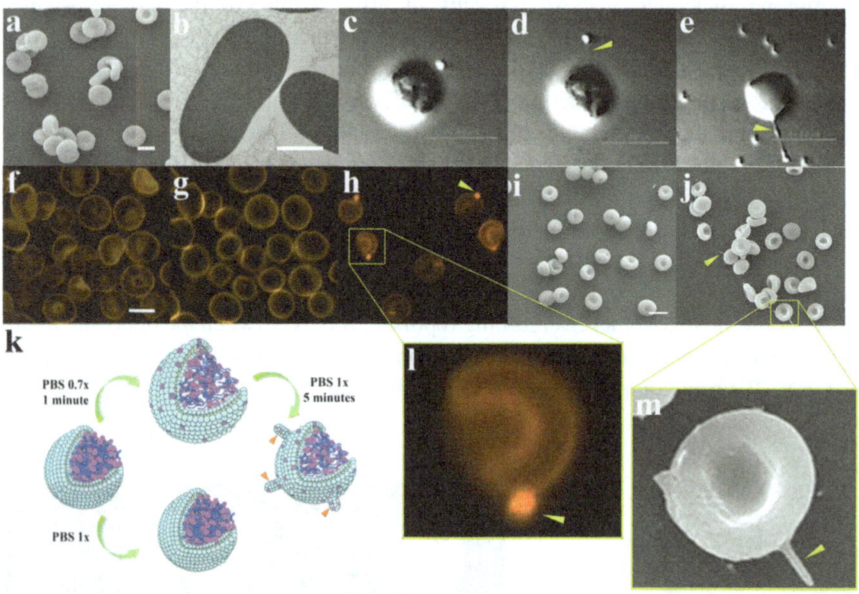

Fig. 8.3 Stretching of the existing cell plasma membrane results in the incorporation of additional membrane to the existing parent membrane [35]. (**a**) Isolated rat RBCs observed by scanning electron microscopy (*Scale, 5* µm) and (**b**) transmission electron microscopy (*Scale, 500* nm), demonstrating the purity of the RBC preparation and the absence of intracellular membrane vesicles. (**c–e**) Light micrographs demonstrating the generation of a membrane tether (yellow arrowheads) formed when an optically trapped 200 nm polystyrene bead adhered to the outer surface of a rat RBC membrane is pulled using the laser tweezers of the photonic force microscope to establish a nearly 6 µm membrane tether (*Scale, 4* µm *in* **c**, **d** *and 8* µm *in* **e**). (**f**) Isolated RBCs in isotonic PBS stained with the R18 Octadecyl Rhodamine B Chloride lipid dye demonstrate plasma membrane staining. (**g**, **i**) Exposure of RBC to hypotonic medium (70% PBS), followed by (**h**, **j**), return to isotonic PBS solution, results in the formation of membrane protrusions, as observed using both light and scanning electron microscopy. (**k**) Schematic illustration of osmotic stretch-induced membrane biogenesis, and (**l**, **m**) high resolution light l, and electron micrograph (**m**) showing protrusions at the RBC membrane (taken from Naik et al., ref. 35). ©Bhanu Jena

the study show that the selective incorporation of proteins and lipids from the cytosol into the cell plasma membrane to generate additional membrane is dictated by the stretch of the existing membrane and its composition [35]. Similar to secretory vesicles, no organelles or membrane-bound vesicles are present within rat red blood cells (RBCs) (Fig. 8.3 from ref. 35), making them an ideal model for determining the incorporation of cytosolic proteins and lipids into the plasma membrane, to generate additional membranes. When RBC plasma membrane is stretched either by pulling on a membrane-tethered polystyrene bead using an optical laser trap (Fig. 8.3c–e) or by rapidly increasing cell volume by subjecting the RBC to a hypotonic buffer solution (Fig. 8.3f–m), excess membrane as tethers and out-pushing or plugs are generated. Such membrane tethers in cells have been extensively investigated, suggesting the contribution of cytoskeletal proteins in their formation in RBCs. Membrane tethers are typically a few tens-of-nanometer thick tubes with lipid bilayer walls, playing critical roles in various physiological processes in the cell such as endoplasmic reticulum and Golgi dynamics, and for the storage of excess lipids. Pulling at 10–15 pN force using laser tweezers on a 200 nm polystyrene bead adhered to rat RBC plasma membrane [35] generates membrane tethers approximately 150 nm in diameter and > 6 μm in length (Fig. 8.3c–e). As illustrated in the transmission electron micrograph in Fig. 8.3b, rat RBCs are devoid of either membrane folds or intracellular vesicles; hence, the rapid generation of excess membrane contributing to membrane tether elongation is the consequence of stretch-induced insertion of cytoplasmic lipids and proteins into the plasma membrane. Similarly, osmotic swelling of isolated RBC induced by incubation in hypotonic (70%) phosphate buffered saline (PBS) solution results in membrane stretching and the addition of cytosolic proteins and lipids to the existing membrane, to form new membrane. When swollen RBCs are returned to isotonic medium (i.e., 100% PBS), membrane pegs or protrusions appear, as observed in both light (Fig. 8.3h, j) and electron micrographs (Fig. 8.3j, m). Pegs represent the excess plasma membrane generated as a result of cell swelling (protein and lipid insertion to provide excess membrane to accommodate the extra volume), followed by shrinking when returned to isotonic medium. Unlike a secretory vesicle, which will lose its spherical form following the loss of intra-vesicular fluid, the RBC, with its well-organized underlying cytoskeleton, is able to recover its original toroidal shape; however, the excess membrane formed as a consequence of incorporated proteins and lipids into the existing cell plasma membrane is therefore pushed outward to form the observed pegs. Mass spectrometry further confirms stretch-induced incorporation of new lipids and proteins into the RBC plasma membrane [35].

MD simulation further supports stretch-induced lipid incorporation into lipid bilayer membranes. NPγT molecular dynamic simulations confirm stretch-induced lipid incorporation into membranes. Figure. 8.4a shows a linear behavior when the imposed tension is plotted as a function of the log of the area per lipid. When the slope of the best-fit line through the data is taken, it produces an area compressibility modulus of $K_A = 234$ mN/m, which is in close agreement with the experimental value of 231 mN/m. To determine the effect of imposed tension on the incorporation of free lipids, a single DPPC lipid fully solvated by water when placed near the

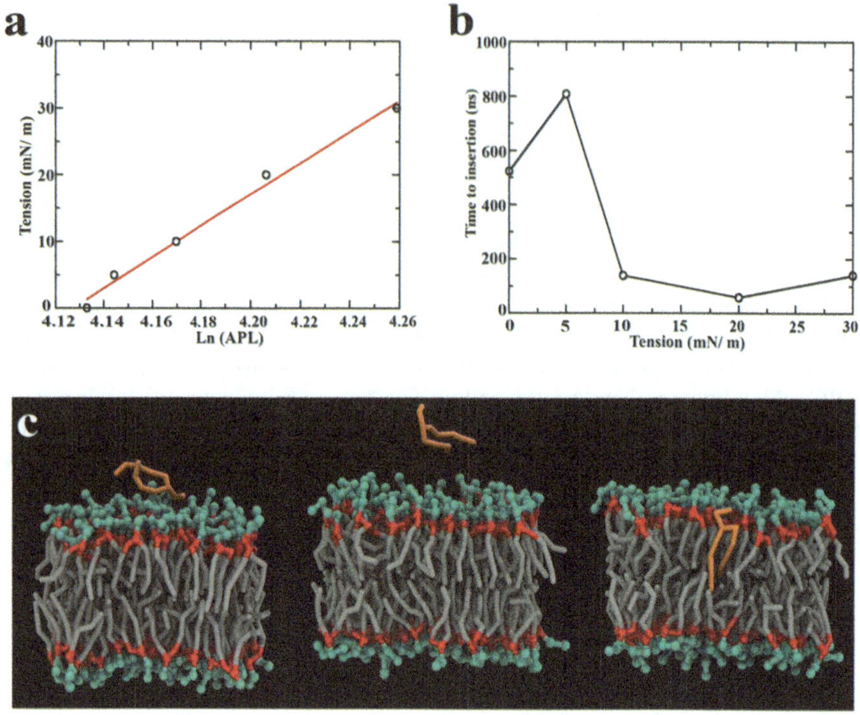

Fig. 8.4 Molecular dynamic simulations demonstrate membrane stretch-induced lipid insertion [35]. (**a**) Applied interfacial tension plotted as a function of the logarithm of the area per lipid. Predictions of molecular dynamics simulation (circles) and linear regression of the simulation data (solid line) are shown. The slope of the curve yields the area compressibility modulus. (**b**) Increased membrane stretch results in the rapid insertion of lipids into the membrane. (**c**) Snapshot from molecular dynamics simulations illustrating the initial configuration (far left), a typical intermediate state (middle), and after lipid incorporation into the bilayer (far right) (taken from Naik et al., ref. 35). ©Bhanu Jena

bilayer surface is inserted into the membrane when the membrane is stretched (Fig. 8.4b, c). Simulations were initiated for a range of interfacial tensions from 0 to 60 dyne/cm. For values of the interfacial tension between 10 and 50 dynes/cm, the free lipid became incorporated into the bilayer within 50–140 ns (Fig. 8.4b). For simulations performed at 0 tension and 5 dynes/cm, incorporation of the free lipid in the bilayer had not occurred until after 500 ns. These results support the premise that application of tension to lipid bilayers enhances significantly the kinetics of free lipid incorporation into bilayer membranes.

The composition of existing membrane dictates the incorporation of specific proteins and lipids into the membrane in membrane biogenesis. In addition to the cell plasma membrane, intracellular organelles such as the nucleus, Golgi apparatus, lysosome, or the endoplasmic reticulum are all encased by membrane, the compositions of which are different and very tightly regulated [44]. The generation of additional membrane must therefore be strictly governed by the composition of the parent membrane in addition to being induced by stretch. This hypothesis has been tested using live Min6 cells [35], an insulin-secreting mouse insulinoma that mimics pancreatic beta cells. Cells incubated with increasing concentrations of methyl beta-cyclodextrin (M-βCD) to deplete cholesterol from the cell plasma membrane demonstrate loss in fluorescent phosphatidylserine (PS) uptake by the plasma membrane, while the uptake of phosphatidylethanolamine (PE) remains unchanged, demonstrating selectivity in lipid uptake by the cell plasma membrane is dictated by the chemistry of parent membrane. Selectivity in lipid species uptake is an interesting observation since it has been previously reported that cholesterol maintenance in the cytoplasmic leaflet of the plasma membrane requires the retention of PS [45]. The reduction of PS uptake upon depletion of cholesterol from the membrane could be due to cholesterol's known participation in cholesterol-mediated lipid–lipid phase separation in the formation of membrane rafts. Lipid phase separations are known to be caused by distinct PS lipid-CHOL versus PG lipid-CHOL interactions. Similarly, PS is known to be confined to the inner leaflet of the cell membrane by the ATP-dependent flippase enzyme ATP11C, counteracting the activity of ATP-independent scramblase. Therefore, when cells are depleted of cholesterol, PS quickly appears at the outer surface, implying that cholesterol acts in the cell as a powerful scramblase inhibitor. The critical role of lipids in the presence and distribution of plasma membrane proteins is also well documented, for example, SNARE proteins are found to be embedded in cholesterol-rich lipid domains [29]. The impact of the loss of cholesterol on plasma membrane proteins such as the presence and distribution of syntaxin, SNAP25, Gαi3, and v-ATPase has also been reported [35]. Extraction of cholesterol from the Min6 cells therefore was found to alter the distribution of SNAP-25 and Gαi3 at the cell plasma membrane and further demonstrated that while there is a loss of syntaxin and SNAP25 at the cell plasma membrane, a gain in the relative amounts of Gαi3 and v-ATPase immunoreactivity was observed. It is interesting to note that both non-membrane and membrane proteins such as hemoglobin, carbonic anhydrase, clathrin, flotillin-2, lin-54 homolog, and TLR4 interactor are found associated with the plasma membrane following RBC stretch [35]. The interaction of hemoglobin with both cytosolic and membrane proteins and lipids have been previously reported, and similarly, earlier studies also show the association of some plasma membrane bicarbonate transporters with carbonic anhydrase enzymes to form bicarbonate transport metabolon to facilitate metabolic CO(2)-HCO(3)(−) conversions and coupled HCO(3)(−) transport. Furthermore, the detection of membrane proteins such as clathrin, flotillin-2, lin-54 homolog, and TLR4 interactor following stretch is a consequence of their selective and directed incorporation into the cell plasma membrane. The results further demonstrate the critical role of membrane cholesterol in maintaining the concentration and distribution of t-SNAREs at the cell plasma membrane. These observations shed light on the establishment of directed higher order structures in live cells and on a number of fundamental and intriguing cellular processes. We can now appreciate how membrane-bound secretory vesicles undergo rapid volume increase required for membrane fusion and the regulated release of intra-vesicular contents during secretion, and how additional cellular membrane is generated at various domains required for cell growth and division among others.

Assembly of calcium between opposing membranes leads to bridging of the oxygens of phospholipid head groups, promoting membrane fusion

Calcium is critical to SNARE-induced membrane fusion. Low fusion rates (τ = 16 min) between t- and v-SNARE-reconstituted liposomes in the absence of Ca^{2+} have been demonstrated [36]. In contrast, t-SNARE and v-SNARE liposomes in Ca^{2+} drive vesicle fusion on a near physiological time-scale (τ ~10 s) [36]. In cells, SNAREs bring opposing bilayers closer to within a distance of 2–3 Å [36].

Calcium ion exists in its hydrated state within cells. The properties of hydrated calcium have been extensively examined using X-ray diffraction, neutron scattering, in combination with molecular dynamics (MD) simulations [37–39]. The MD simulations include three-body corrections compared with ab initio quantum mechanics/molecular mechanics molecular dynamics simulations. First principle molecular dynamics has also been used to investigate the structural, vibrational, and energetic properties of $[Ca(H_2O)_n]^{2+}$ clusters and the hydration shell of the calcium ion. These studies show that hydrated calcium $[Ca(H_2O)_n]^{2+}$ has more than one shell around the Ca^{2+}, with the first hydration shell having six water molecules arranged in an octahedral conformation [37]. Light scattering and X-ray diffraction studies of SNARE-reconstituted liposomes demonstrate that fusion proceeds only when Ca^{2+} ions are available between the t- and v-SNARE-opposed membranes [36, 40]. X-ray diffraction studies show that in the absence of Ca^{2+} a diffuse and asymmetric diffractogram characteristic of short range ordering in a liquid system is observed [38]. In contrast, in the presence of Ca^{2+} t-SNARE and v-SNARE proteoliposomes exhibit a more structured diffractogram, with approximately a 12% increase in X-ray scattering intensity, suggesting an increase in the number of contacts between opposing lipid membranes, established presumably through calcium-phosphate bridges, as previously suggested [36, 39, 40]. The ordering effect of Ca^{2+} on inter-bilayer contacts observed in X-ray studies [36] is in good agreement with light, AFM, and spectroscopic studies, suggesting close apposition of PO-lipid head groups in the presence of Ca^{2+}, followed by formation of Ca^{2+}-PO bridges between the adjacent bilayers [36, 40]. X-ray diffraction studies show that the effect of Ca^{2+} on bilayers orientation and inter-bilayer contacts is most prominent in the area of 3 Å, with additional appearance of a new peak at position 2.8 Å, both of which are within the ionic radius of Ca^{2+} [36]. These studies further suggest that the ionic radius of Ca^{2+} may make it an ideal player in the membrane fusion reaction. Hydrated calcium $[Ca(H_2O)_n]^{2+}$, however, with a hydration shell having six water molecules and measuring ~6 Å would be excluded from the t-/v-SNARE apposed inter-bilayer space; hence, calcium has to be present in the buffer solution when t-SNARE vesicles and v-SNARE vesicles are added. Indeed, studies demonstrate that if t- and v-SNARE vesicles are allowed to mix in a calcium-free buffer, there is no fusion following post-addition of calcium [40].

MD simulation further confirms the bridging of tightly opposed bilayers by Ca^{2+} leading to the release of water from the hydrated Ca^{2+} ion and the co-ordinated waters from the oxygens of the phospholipid head groups, leading to bilayer desta-

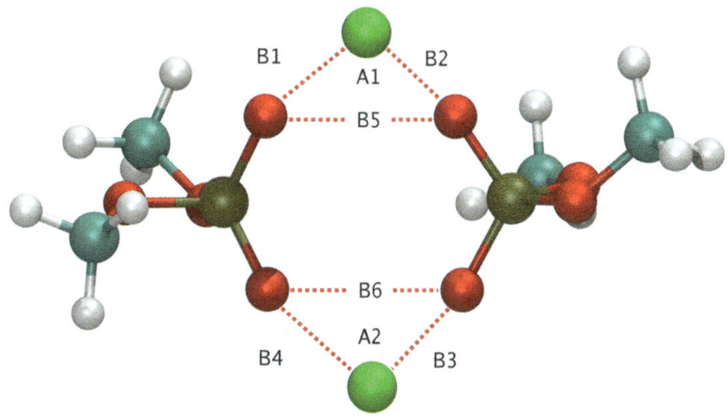

Fig. 8.5 Ca^{2+}-DMP^{-} ring complex observed during molecular dynamics simulations. Atoms are colored as follows: Carbon (cyan), hydrogen (white), oxygen (red), phosphorous (gold), and Ca^{2+} (green). B1 = 2.146 Å, B2 = 2.145 Å, B3 = 2.140 Å, B4 = 2.145 Å, B5 = 2.92 Å, B6 = 2.92 Å, A1 = 85.79°, A2 = 85.07° [41]. ©Bhanu Jena

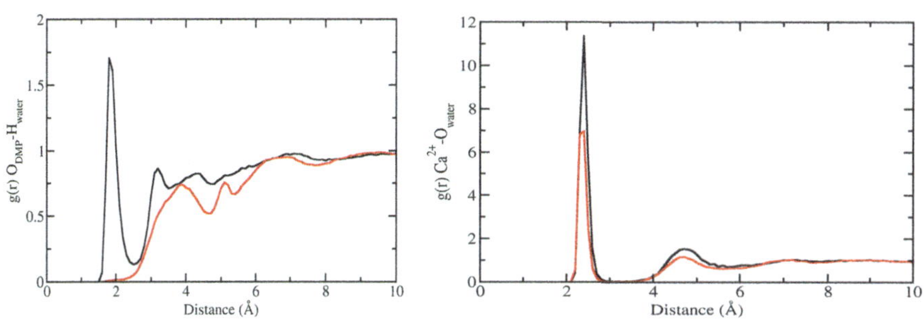

Fig. 8.6 Radial distribution functions for water interacting with the DMP^{-} and Ca^{2+}. (**Left**) Hydrogen atoms of water interacting with anionic DMP^{-} oxygens. Black curve corresponds to DMP- in free solution; red curve corresponds to DMP- bound to Ca^{2+}. (**Right**) Oxygen of water interacting with Ca^{2+}. Black curve represents Ca^{2+} unbound to DMP-; red curve represents Ca^{2+} bound to DMP^{-} [41]. ©Bhanu Jena

bilization and membrane fusion [41]. Using MD simulations in the isobaric–isothermal ensemble to determine whether Ca^{2+} was capable of bridging opposing phospholipid head groups in the early stages of the membrane fusion process demonstrates indeed this to be the case (Figs. 8.5 and 8.6). Furthermore, the distance between the oxygen atoms of the opposing PO-lipid head groups bridged by calcium [41] is in agreement with the 2.8 Å distance reported using X-ray diffraction measurements [36].

Results from studies in the past few decades, utilizing a combination of multimodal imaging approaches, spectroscopic studies, and chemical synthesis, collectively provide greater insights into the assembly and disassembly of cellular higher order structures and attest to chirality and enthalpy as critical drivers in the assembly and function of higher ordered supramolecular structures in cells. Since supramolecular structures at equilibrium have lower Gibbs free energy, they are phase separated.

References

1. Zwang, T. J., Hurlimann, S., Hill, M. G., & Barton, J. K. (2016). Helix-dependent spin filtering through the DNA duplex. *Journal of the American Chemical Society, 138*, 15551–15554.
2. Kumar, A., Capua, E., Kesharwani, M. K., Martin, J. M. L., Sitbon, E., Waldeck, D. H., & Naaman, R. (2017). Chirality-induced spin polarization places symmetry constraints on biomolecular interactions. *Proceedings of the National Academy of Sciences of the United States of America, 114*, 2474–2478.
3. McDormott, M. L., Vanselous, H., Corcelli, S. A., & Petersen, P. B. (2017). DNA's chiral spine of hydration. *ACS Central Science, 3*, 708–714.
4. Joyce, G. F., Visser, G. M., van Boeckel, C. A. A., van Boom, J. H., Orgel, L., & van Westrenen, J. (1984). Chiral selection in poly(C)-directed synthesis of oligo(G). *Nature, 310*, 602–604.
5. Avetisov, V. A., & Goldanskii, V. I. (1991). Homochirality and stereospecific activity: Evolutionary aspects. *Biosystems, 25*, 141–149.
6. Blackmond, G. G. (2010). The origin of biological homochirality. *Cold Spring Harbor Perspectives in Biology, 2*, a002147. Edited by David Deamer and Jack W. Szostak.
7. Park, J. T., & Strominger, J. I. (1957). Mode of action of penicillin. *Science, 125*, 99–101.
8. Hancock, R. (1960). The amino acid composition of the protein and cell wall of *Staphylococcus aureus*. *Biochimica et Biophysica Acta, 37*, 42–46.
9. Lam, H., Oh, D.-C., Cava, F., Takacs, C. N., Clardy, J., de Pedro, M. A., et al. (2009). D-Amino acids govern stationary phase cell wall remodeling in bacteria. *Science, 325*, 1552–1555.
10. Kolodkin-Gal, I., Romero, D., Cao, S., Clardy, J., Kolter, R., & Losick, R. (2010). D-amino acids trigger biofilm disassembly. *Science, 328*, 627–629.
11. Kepert, I., Fonseca, J., Müller, C., Milger, K., Hochwind, K., Kostric, M., et al. (2017). D-tryptophan from probiotic bacteria influences the gut microbiome and allergic airway disease. *The Journal of Allergy and Clinical Immunology, 139*, 1525–1535.
12. Epand, R. M., Rychnovsky, S. D., & Belani, J. D. (2005). Role of chirality in peptide-induced formation of cholesterol-rich domains. *The Biochemical Journal, 390*, 541–548.
13. Wezenberg, S. J., & Feringa, B. L. (2018). Supramolecularly directed rotary motion in a photosensitive receptor. *Nature Communications, 9*, 1984–1990.
14. Leigh, D. A., Wong, J. K. Y., Dehes, F., & Zerbetto, F. (2003). Unidirectional rotation in a mechanically interlocked molecular rotor. *Nature, 424*, 174–179.
15. Hernández, J. V., Kay, E. R., & Leigh, D. A. (2004). A reversible synthetic rotary molecular motor. *Science, 306*, 1532–1537.
16. Wilson, M. R., Sola, J., Carlone, A., Goldup, S. M., Lebrasseur, N., & Leigh, D. A. (2016). An autonomous chemically fuelled small-molecule motor. *Nature, 534*, 235–240.
17. Astumian, R. D. (2016). Optical vs. chemical driving for molecular machines. *Faraday Discussions, 195*, 583–597.
18. Pezzato, C., Cheng, C., Stoddart, J. F., & Astumian, R. D. (2017). Mastering the non- equilibrium assembly and operation of molecular machines. *Chemical Society Reviews, 46*, 5491–5507.
19. Nesci, S., Trombetti, F., Ventrella, V., & Pagliarani, A. (2015). Opposite rotation directions in the synthesis and hydrolysis of ATP by ATP synthase: Hints from a subunit asymmetry. *The Journal of Membrane Biology, 248*, 163–169.

20. Harkness, R. W., Avakyan, N., Sleiman, H. F., & Mittermaier, A. K. (2018). Mapping the energy landscapes of supramolecular assembly by thermal hysteresis. *Nature Communications, 9*, 3152–3161.
21. Mergny, J. L., & Lacroix, L. (2003). Analysis of thermal melting curves. *Oligonucleotides, 13*, 515–537.
22. Small, D. M., & Shipley, G. G. (1974). Physical-chemical basis of lipid deposition in atherosclerosis. *Science, 185*, 222–229.
23. Mahamid, J., Tegunov, D., Maiser, A., Arnold, J., Leonhardt, H., Plitzko, J. M., & Baumeister, W. (2019). Liquid-crystalline phase transitions in lipid droplets are related to cellular states and specific organelle association. *Proceedings of the National Academy of Sciences of the United States of America, 116*, 16866–16871.
24. Baumgart, T., Hammond, A. T., Sengupta, P., Hess, S. T., Holowka, D. A., Baird, B. A., & Webb, W. W. (2007). Large-scale fluid/fluid phase separation of proteins and lipids in giant plasma membrane vesicles. *Proceedings of the National Academy of Sciences of the United States of America, 104*, 3165–3170.
25. Walter, H., & Brooks, D. E. (1995). Phase separation in cytoplasm, due to macromolecular crowding, is the basis for microcompartmentation. *FEBS Letters, 361*, 135–139.
26. Humphrey, D., Duggan, C., Saha, D., Smith, D., & Käs, J. (2002). Active fluidization of polymer networks through molecular motors. *Nature, 416*, 413–416.
27. Brangwynne, C. P., Eckmann, C. R., Courson, D. S., Rybarska, A., Hoege, C., Gharakhani, J., et al. (2009). Germline P granules are liquid droplets that localize by controlled dissolution/condensation. *Science, 324*, 1729–1732.
28. Simons, K., & Ikonen, E. (1997). Functional rafts in cell membranes. *Nature, 387*, 569–572.
29. Wang, S., Lee, J. S., Bishop, N., Jeremic, A., Cho, W. J., Chen, X., Mao, G., Taatjes, D. J., & Jena, B. P. (2012). 3D organization and function of the cell: Golgi budding and vesicle biogenesis to docking at the porosome complex. *Histochemistry and Cell Biology, 137*, 703–718.
30. Cho, S.-J., Kelly, M., Rognlien, K. T., Cho, J. A., Horber, J. K. H., & Jena, B. P. (2002). SNAREs in opposing bilayers interact in a circular array to form conducting pores. *Biophysical Journal, 83*, 2522–2527.
31. Cho, W.-J., Jeremic, A., & Jena, B. P. (2005). Size of supramolecular SNARE complex: Membrane-directed self-assembly. *Journal of the American Chemical Society, 127*, 10156–10157.
32. Jeremic, A., Quinn, A. S., Cho, W.-J., Taatjes, D. J., & Jena, B. P. (2006). Energy-dependent disassembly of self-assembled SNARE complex: Observation at nanometer resolution using atomic force microscopy. *Journal of the American Chemical Society, 128*, 26–27.
33. Shin, L., Cho, W.-J., Cook, J., Stemmler, T., & Jena, B. P. (2010). Membrane lipids influence protein complex assembly-disassembly. *Journal of the American Chemical Society, 132*, 5596–5597.
34. Cho, W. J., Lee, J.-S., Ren, G., Zhang, L., Shin, L., Manke, C. W., Potoff, J., Kotaria, N., Zhvania, M. G., & Jena, B. P. (2011). Membrane-directed molecular assembly of the neuronal SNARE complex. *Journal of Cellular and Molecular Medicine, 15*, 31–37.
35. Naik, A. R., Kuhn, E. R., Lewis, K. T., Kokotovich, K. M., Maddipati, K. R., Chen, X., Hörber, J. H. K., Taatjes, D. J., Potoff, J. J., & Jena, B. P. (2019). Self-assembly and biogenesis of the cellular membrane are dictated by membrane stretch and composition. *The Journal of Physical Chemistry. B, 123*, 6997–7005.
36. Jeremic, A., Kelly, M., Cho, J.-H., Cho, S.-J., Horber, J. K. H., & Jena, B. P. (2004). Calcium drives fusion of SNARE-apposed bilayers. *Cell Biology International, 28*, 19–31.
37. Bako, I., Hutter, J., & Palinkas, G. (2002). Car-Parrinello molecular dynamics simulation of the hydrated calcium ion. *The Journal of Chemical Physics, 117*, 9838–9843.
38. McIntosh, T. J. (2000). Short-range interactions between lipid bilayers measured by X-ray diffraction. *Current Opinion in Structural Biology, 10*, 481–485.

39. Portis, A., Newton, C., Pangborn, W., & Papahadjopoulos, D. (1979). Studies on the mechanism of membrane fusion: Evidence for an intermembrane Ca^{2+} phospholipid complex, synergism with Mg^{2+}, and inhibition by spectrin. *Biochemistry, 18*, 780–790.
40. Jeremic, A., Cho, W. J., & Jena, B. P. (2004). Membrane fusion: What may transpire at the atomic level. *Journal of Biological Physics and Chemistry, 4*, 139–142.
41. Potoff, J. J., Issa, Z., Manke, C. W., Jr., & Jena, B. P. (2008). Ca^{2+}-Dimethylphosphate complex formation: Providing insight into Ca^{2+} mediated local dehydration and membrane fusion in cells. *Cell Biology International, 32*, 361–366.
42. Weber, T., Zemelman, B. V., McNew, J. A., Westerman, B., Gmachi, M., Parlati, F., Söllner, T. H., & Rothman, J. E. (1998). SNAREpins: Minimal machinery for membrane fusion. *Cell, 92*, 759–772.
43. Rothman, J. E. (1994). Mechanisms of intracellular protein transport. *Nature, 372*, 55–63.
44. Lewis, K. T., Maddipati, K. R., Naik, A. R., & Jena, B. P. (2017). Unique lipid chemistry of synaptic vesicle and synaptosome membrane revealed using mass spectrometry. *ACS Chemical Neuroscience, 8*, 1163–1169.
45. Bogan, J. S., Xu, Y., & Hao, M. (2012). Cholesterol accumulation increases insulin granule size and impairs membrane trafficking. *Traffic (Copenhagen, Denmark), 13*, 1466–1480.

Index

B. P. Jena, *Cellular Nanomachines*,
https://doi.org/10.1007/978-3-030-44496-9

The manufacturer's authorised representative in the EU is Springer Nature Customer Service Centre GmbH, Europaplatz 3, 69115 Heidelberg, Germany. If you have any concerns regarding our products, please contact ProductSafety@springernature.com

Printed and bound by CPI Group (UK) Ltd, Croydon, CR0 4YY

28/09/2026

02224089-0015